29 95

Spontaneous Phenomena

A Mathematical Analysis

Spontaneous Phenomena

A Mathematical Analysis

Flemming Topsøe

Københavns Universitet
Matematisk Institut
København, Denmark

Translated by John Stillwell

ACADEMIC PRESS, INC.
Harcourt Brace Jovanovich, Publishers
Boston San Diego New York
Berkeley London Sydney
Tokyo Toronto

This book is printed on acid-free paper. ♾

ACADEMIC PRESS, INC.
1250 Sixth Avenue, San Diego, CA 92101

United Kingdom Edition published by
ACADEMIC PRESS LIMITED
24–28 Oval Road, London NW1 7DX

Library of Congress Cataloging-in-Publication Data
Topsøe, Flemming.
[Spontane fænomener. English]
Spontaneous phenomena : a mathematical analysis / Flemming Topsøe
: translated by John Stillwell.
p. cm.
Translation of: Spontane fænomener.
Bibliography: p.
Includes index.
ISBN 0-12-695175-6 (alk. paper)
1. Mathematical physics. 2. Mathematical analysis. I. Title.
QC20.T6613 1989
530.1'5 — dc20 89-15017
CIP

First published in the Danish Language under the title
"Spontane Fænomener," and copyrighted in
1983 by Nyt Nordisk Forlag Arnold Busck, A/S, København, Denmark.

Printed in the United States of America
89 90 91 92 9 8 7 6 5 4 3 2 1

CONTENTS

Preface and Introduction *vii*
List of Symbols *xi*

1. The Problem 1
2. Thoughts on Model Building 5
3. On Stochastic Models 9
4. Time Invariance 17
5. Independence 21
6. Intermezzo on the Binomial Distribution 23
7. On the Physical Background to the Assumptions 27
8. The Intensity 29
9. Mathematics, at last! 31
10. Confrontation with Reality 39
11. Critique of the Mathematics 43
12. Critique of the Physics 47
13. Other Applications of the Model 51
14. The Poisson Approximation to the Binomial Distribution 53
15. Spatially Uniform Distribution, Point Processes 59
16. A Detailed Analysis of Actual Observations 65

17. The chi-square Test for Goodness of Fit 81
18. The Historical Perspective 93
Exercises 113
Examples for Further Investigation 141
Programs 161

References 177
Index 179

PREFACE AND INTRODUCTION

First, let me tell you how this book came about. The background was a series of discussions in the Danish *gymnasium*[1] aiming at introducing certain *aspects* in the teaching of mathematics. The aspects concerned the *historical perspective*, *modelbuilding* and the *inner nature* of mathematics. Furthermore, there was a pronounced desire to introduce the students to authentic applications of mathematics. And, of course, the exploitation of electronic data processing should, at best, be integrated in the instruction. The tendencies, not much different from what went on in many other countries, were, to a certain extent, a result of the spirit from '68. Many a wise word were said and written in this context. But the discussion could, undoubtedly, appear somewhat academic. All these words *about* what to do...

With this background I felt it natural to seize the pen with the aim of indicating a way to handle the problems through a discussion, and here I mean a mathematical discussion, of a concrete example. A case story in mathematical modelling if you wish. The point of departure had to be a relatively simple problem. I chose to look at radioactivity. Due to the students' familiarity with this phenomenon, explained in the physics classroom and often referred to in one way or another in the press, I gathered that it would be relatively easy to understand what the problem was about (cf. Chapter 1).

[1]The gymnasium gives training at the level of the last pre-college years and the first year at college.

The study of radioactivity gives rise to reflections on the notion of spontaneity. Probability theory can be seen in a new light, not only as an effective tool, but more so as a principal necessity in the description of nature. Thus the subject has fascinating philosophical overtones. Various discussions could emerge - say, on the existance of a free will! Not that I embark on such discussions in the text. After all, a book is not what is written but the opportunities it offers.

It was also important for the choice of subject that, as it were, all the different points of view which were demanded for in the debate referred to above, could be integrated, at least touched upon, in the text.

It has to be realized that the result of my endeavours - the book in hand - is in fact a contribution to this debate. As such I have taken certain liberties. The exposition is not homogeneous as in a textbook, the mathematics involved goes beyond what is usually taught at this level and for certain parts, this especially applies to the exercises, the demands on the ability of the reader are quite high. Furthermore, I have attempted a rather free style of writing in order to convey attitudes towards the subject. This is done in some comprehensive discussions as well as in the formulation of certain *theses*. For sure, these theses are not all equally profound, but anyhow, they may give the incitement needed for the readers further considerations and reflections.

Mathematics does not thrive in a vacuum but demands a manysided cooperation involving personal attitude and consideration of the exterior, material world. In such an extended context, mathematics is not absolute, not just a yes/no-subject. The essence is knowledge, but knowledge with more facets than by the study of isolated abstract mathematics, and knowledge which does not come just by the dissection of epsilons and deltas. Hopefully, the book mirrors this conception and argues for it.

Considering the aim of the book, it is natural to claim that it points to new and practicable paths for mathematical instruction at pre-college and early college levels. Some may, however, find that rather, the text underlines that sensible attitudes to mathematical material requires a high degree of maturity, and that the treatment of authentic applications of mathematics often becomes quite sophisticated and technically involved - unless you satisfy yourself with superficial simplifications.

Concerning the use of the book, I recommend that it is read under skilled guidance. This applies, in particular, if you want to study the exercises. Under guidance, it is also possible to make a shortcut in arriving at the model for radioactive decay (cf. Exercise 16). The more elaborate considerations of the main text are, however, an advantage later on (Chapters 14 and 15). The material treated is well suited for mathematics classes. However, interdisciplinary courses combining mathematics and physics should also be attractive. In this connection it may be noted that the book contains detailed directions (Chapter 16) for the performance of experiments with equipment which is likely to be available at every college and at most high schools. There is a detailed discussion of the handling of data arising from such experiments, especially concerning the complications which arise due to the inevitable imperfections of the measuring apparatus (problems of paralysis). In order to make a control based on statistic principles possible, one chapter is devoted to the chi- squared test for goodness of fit.

The main text ends with a historical chapter giving a picture of the development of the relevant mathematics and physics.

A substantial part of the book follows after the main text. First comes a section containing 46 exercises. They vary widely in difficulty as well as in the aim - from very theoretical ones (which many will surely skip) to exercises dealing with statistical principles and other matters directed towards applications. Perhaps, it is proper to single out Exercise 38 as this exercise leads to a model for radioactive decay which is more accurate than the one discussed in the main text.

Then follows a section which is simply a collection of authentic data, essentially without any hints as to how one might handle the data. This is left entirely to the reader. The data concerns various phenomena within biology, medicin, statistics of accidents, determination of age, traffic analysis and other fields, and they have all been collected out of interest for the phenomena they concern and not as an academic exercise giving rise to a mathematical analysis.

At last you will find some EDP-programs which may greatly facilitate the handling of data given in the book or elsewhere. The programs are reasonably well documented. Some readers may prefer to use the documentation as starting point for writing their own programs, tailor made to their needs and personal taste.

I should be pleased to see if the book will also be found of use in teaching at the more advanced university level. Then the main text can be dealt with quite quickly, or skipped altogether, and one can concentrate on the exercises and the collection of authentic data. A problem seminar would be quite ideal for these activities. I also believe that many will find it rewarding to carry out experiments following the guidance given in Chapter 16 and in some of the exercises. In fact, this leads to more accurate determinations than you find in the standard physics textbooks.

Lastly, I should like to mention that I have recieved valuable advice concerning the preparation of the danish edition from the following persons, mainly colleagues at the University of Copenhagen: Erik Sparre Andersen, Lissen Haugwitz, Michael Hedegaard, Søren Johansen, Malte Olsen and Erik Rüdinger. For the translation into English (except this foreword and the section with programs), I want to thank John Stillwell, and for technical assistance I thank Klaus Harbo, Leif Mejlbro and Anders Thorup.

Flemming Topsøe
Copenhagen, July 11, 1989

List of Symbols

t_n	2	v_n	2
$N(t)$	2	$P(X \in A)$	14
$F(x)$	15	$E(X)$	15
$N(I)$	20	$P(A\|B)$	22
$\binom{n}{k}$	24	λ	29
μ	30, 119	U	39
N_k	39	V	39
λ_{obs}	40	T	40
N_k^*	40	p_{det}	49
λ_{det}	49	$N_{det}(t)$	49
h	49	$N_{total}(t)$	69
λ_{reg}	71	$P_{\lambda,h}$	73
$N_{reg}(t)$	73	$L(\cdot)$	74, 122
χ_n^2	82	d	87
D	88	SL	88
σ^2	119	1_H	120
$\hat{\lambda}$	122	$\bar{x}$	129
s^2	129	q^2	129
W	130	$N_{unreg}(t)$	131
$T_{1/2}$	135	λ_k	136
$x_k(t)$	136	T_k	136
$A_k(t)$	136		

1. The Problem

Observing a radioactive preparation with the help of a Geiger counter, one can record the moments at which certain physical events occur: A count is made when one hears a sharp sound, a "click".

This is due to the phenomenon of atomic transformation. In general, we shall use a neutral terminology, referring to such transformations as *events*. Now and then, however, we shall use the more specific term *decay*. This word is used here as a generic term, both in the case of a passage from an excited state with high energy to a state of lower energy (and subsequent emission of a γ-quantum), and in the case of a proper nuclear decay, as, e.g., for α- or β-active preparations.

In practice it is difficult to determine the precise moment of decay. This is especially true when there are many decays per unit time (the intensity is high). For our more theoretical considerations this will play no role, but we shall take it into account when we check the theory later (§10).

It is important to remark that we confine ourselves to radioactive processes where the half life of the preparation is so long that it is impossible to detect any change in the preparation in the course of observation. We can therefore say that the intensity is constant with time.

The results of an experiment can be given on a time axis, with time 0 corresponding to the moment at which we begin observations. We mark the time axis with a cross whenever an event is registered (a click is heard).

We let t_1 denote the moment of the first event, t_2 the moment of the second event and so on. We call the t's *arrival times* (arrival of the physical events, or "clicks").

We denote by v's the *waiting times*, i.e., v_1 is the waiting time for the first event, v_2 is the waiting time for the next event, etc.

Finally we introduce some quantities, N's, which give the *number of events* up to a given moment. More precisely, for each $t > 0$ we set $N(t)$ = the number of events up to time t.

Figure 1 shows the arrival times, waiting times and the number of events for a certain observation.

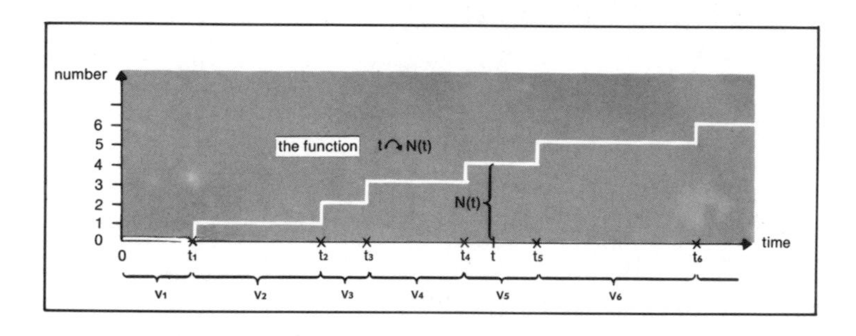

Fig. 1.

An observation (of infinite length) is uniquely determined by one of the following three sets of data: $(t_n)_{n\geq 1}$, $(v_n)_{n\geq 1}$ or $(N(t))_{t>0}$.

It is not crucial how $N(t)$ is defined when t is an arrival point. We can readily choose $N(t)$ to be the number of events in the interval $(0, t] = \{s \mid 0 < s \leq t\}$. This amounts to giving the function $t \mapsto N(t)$ the "highest" value at jump points. We set $N(0) = 0$. In order to take account of all possibilities, we agree that if it should happen that an event occurs exactly at the beginning of the observation, at time 0, it will be ignored. Thus we shall always have $t_1 > 0$. Moreover, one can insist on beginning the observation exactly at the occurrence of an event, in order to put the first waiting time on the same footing as the other waiting times. However, it should now be mentioned that for the phenomena we consider it does not matter when the observation starts. We have:

(1) $$v_n = t_n - t_{n+1}; \qquad n = 1, 2, \ldots$$

(2) $$t_n = v_1 + v_2 + \cdots + v_n; \qquad n = 1, 2, \ldots$$

(here we set $t_0 = 0$ so that (1) holds for $n = 1$).

The connection with the N's is not so nice, purely formally, but it is clear from the figure. Thus t_n is the moment when $N(t)$ jumps from $n-1$ to n, and v_n is the length of the interval in which $N(t)$ has the value $n-1$.

The physicist has discovered that for a wide range of preparations he obtains experimental results of the same irregular appearance as that shown by Figure 1. He finds that this expresses something essential in the phenomenon, and hence in radioactive processes, and therefore wishes to be able to "explain" such sets of observations by physical considerations.

It lies nearby to conclude that the phenomenon is *nondeterministic*, i.e. unpredictable. Nevertheless, the physicist feels instinctively that there is a certain regularity present – that not *every* series of possible arrival times we could write down would in fact be a conceivable result of observation of a radioactive preparation.

It is at this moment that the physicist turns to us mathematicians for help. A set of observations is indeed a series of numbers – e.g., the arrival times $t_1, t_2, \dots$ or the waiting times $v_1, v_2, \dots$ – and it is not wrong to expect that mathematicians will have some ability to handle and "explain" numbers. So here is the problem we are presented with:

PRESENTATION OF THE PROBLEM. *"I am a physicist, interested in radioactivity. My observations exhibit a highly irregular character, as you can see, but of course they must be regular nevertheless. Find the regularity. Bring order to this disorder!"*

Who would expect that the regularity in fact found would only be explicable on the basis of the phenomenon's being unpredictable in principle? An almost paradoxical assertion yet, under the closer examination we are about to give, it will prove to be reasonable.

2. Thoughts on Model Building

We remain under an obligation to stick with the physicist; several matters must be clarified before we understand what it is he will have us do. Certainly we perceive what the problem is, but if we merely try our own hand at it we risk arriving at a solution that corresponds to something not interesting to the physicist, or which is built upon assumptions he cannot accept.

As a first question, we may ask the physicist why there "naturally" must be a certain regularity present. And I believe we shall get the answer that the physical laws are invariable with time and, in addition, the preparations consist of an enormous number of atoms compared with the number of events observed, so that the behaviour of the system will not be affected at all by the events that take place.

There is only one way in which one can obtain a better understanding of such considerations and thereby enter into the spirit of the problem. This is by becoming better acquainted with the physical background. We must "pump" the physicist for information. If we are energetic enough, perhaps we can get him to develop a large part of atomic theory for us. We may also learn about radioactive preparations as they are found in minerals, certain elements and salts, among other things. Perhaps the account will be spiced with historical remarks.

After making the effort to get to the underlying physics (!), we arrive at the stage of constructing a mathematical model. We have a wealth of information at our command and, as a guiding principle in what follows, we set up the following:

THESIS 1. *In model building, all surplus baggage should be thrown overboard.*[2]

We can also say that we seek to *compress* or *reduce* data.

At this stage we dismiss the physicist. He must leave together with all his detailed knowledge. We shall only keep the mathematically relevant information, and seek to summarise its essential characteristics.

I shall now run a certain risk of boring readers with some extra considerations concerning the relation between real phenomena and mathematical models. First the obvious:

THESIS 2. *Not every phenomenon has a mathematical model.*

The pure *description* of a phenomenon, even one without a mathematical model can, however, be more or less mathematical. The next thesis is perhaps not quite so obvious:

THESIS 3. *No phenomenon is mathematics.*

If we think of the concrete phenomenon concerning radioactivity, we probably see well enough what this means. Indeed, let us suppose that we *have* set up a mathematical model. For a physicist, the physics is not by any means replaced by the mathematical model. If anything, one can say: on the contrary! With the mathematical model in hand, he has overcome a difficulty which obstructed his study, so that with this mathematical tool at his command he can really throw himself properly into the subject of interest, namely radioactivity.

Consequently, it is not the primary task of the mathematician, in his relation to applications, to be a missionary for his subject. But it is reasonable to demand that users have a certain knowledge of mathematics. For if the user is to use the mathematical model with sufficient ease that the essential steps are clear to him – he must for example not get lost in mathematical details – it is necessary for him to have a good knowledge of the mathematics involved. However, it is not necessary for him to have the same level of knowledge and understanding as a mathematician.

THESIS 4. *Each user of mathematics must understand mathematics.*

This thesis has a natural solidarity with the following, which we have expressed already at the beginning of our discussion.

---

[2]Also known as Ockham's razor (Translator's note)

THESIS 5. *Each applied mathematician must have insight into the field of application.*

A natural consequence of the two theses above is that especially fruitful results will be obtained when mathematical ability is combined with insight into the field of application. Two famous examples of this assertion are the works of Archimedes and Newton.

In this book we shall see mathematics primarily in the role of serving a particular field of application. And in many other situations one will see mathematics exclusively as a tool. However, one can certainly cultivate mathematics for its own sake. In order to give the subject of mathematics its due, I therefore urge:

THESIS 6. *There are two reasons for the cultivation of mathematics: as a means of obtaining knowledge in other fields, and as knowledge in itself.*

Unfortunately, not everybody considers the second reason to be a completely valid reason for devoting one's time to mathematics. Reflecting on this can certainly be interesting, but I have already moved too far away from the concrete problem.

3. On Stochastic Models

The word *stochastic* means random.[3] A *stochastic model* is the same as a probability model.

THESIS 7. *Mathematical models can be deterministic or stochastic.*

– And mixtures of the two, of course.

The phenomenon which interests us is "obviously" nondeterministic, so the natural idea is to look for a stochastic model.

There are several ways of viewing stochastic models.

THESIS 8. *Stochastic models reflect human ignorance and impotence; in principle all phenomena are deterministic. It is only our lack of insight and ability which forces us to rely on observations.*

To see the reasonableness of this thesis, one can think, e.g., of flipping a coin. If one had complete information about the coin's position, the forces acting on it, and so on, then in principle (!) it would be possible to foresee whether the outcome would be heads or tails.

As a counter-thesis to Thesis 8 we have the following:

THESIS 9. *Some phenomena are stochastic in principle.*

Personally, I definitely prefer the counter-thesis, but it is undecidable in the nature of things, or even a meaningless question, which of the two theses is true. Moreover, one can go a little further:

---

[3]The word comes from the Greek *stochos* (στόχος) for target. The associated verb *stochazomai* (ό στοχάζομαι) means "I throw at the target."

THESIS 10. *All phenomena are stochastic in principle – it is only our inability to make detailed observations which makes some phenomena appear deterministic.*

Here one can think, e.g., of statistical thermodynamics, where disorder at the microlevel (among the individual atoms of a gas, for example) is consistent with order at the macrolevel (precise values of pressure, temperature, etc.). A central example of this state of affairs, and the one which is our main interest, radioactivity, is treated in Exercise 38.

Regardless of whether no (Thesis 8), some (Thesis 9) or all (Thesis 10) phenomena are stochastic, let us agree that for the problem before us, namely radioactivity, the goal is to set up a stochastic model. Even if this phenomenon should be deterministic in principle, our discussion with the physicist has taught us that a deterministic description is impossible (perhaps even undesirable). The usefulness of a stochastic model must be decided by investigating how well it describes reality; more of this later.

And now for some slightly more concrete considerations concerning stochastic models. When can they be used?

THESIS 11. *Stochastic models can be used to describe non-deterministic phenomena of a repetitive character.*

– preferably they apply to phenomena that can be described by trials that can be repeated many times under uniform conditions and without influencing each other (independent trials).

A thesis closely connected with this is:

THESIS 12. *Probabilities are determined empirically as relative frequencies in a large number of trials. Expectations (mean values) are determined empirically as averages in a large number of trials.*

The attitude expressed by this thesis is upheld in the mathematical world by the *law of large numbers* (see Exercise 15). But the thesis is also expressed in the real world, the world of experience. For it is a fact of experience that relative frequencies stabilise.

Let us think, e.g., of the event of "throwing a six with one die" and suppose that a series of throws is made so that the result of each trial does not influence the outcome of subsequent trials (independence!). The *frequency* of the event is the number of trials that result in a six, and the *relative frequency* is the frequency divided by the total number of trials. The more trials there are in the experiment, the more pronounced it becomes that the relative frequency stabilises around a certain value, which we interpret as the *probability* of the event in question.

The interpretation of probability just sketched is called the *frequency interpretation* and it is this interpretation which – in the opinion of many – is the best starting point for an introduction to probability theory. Even if the reader already knows something of probability theory, we shall elaborate further on this interpretation below.

We first remark that, in the example given, the frequency interpretation is superfluous in a way, since it is "clear" beforehand (*a priori*) that the probability of a six is $\frac{1}{6}$. Certainly, most people will accept that, but what is the real meaning of such a statement? The bare numerical value without any interpretation is scarcely of any value. And I wonder whether most people, after further thought, would not arrive at the frequency interpretation? For example, if it was claimed that the probability of a six was $\frac{1}{2}$, how would one refute it? By making a series of trials, of course! And if the relative frequency turns out to be far from $\frac{1}{2}$ (and close to $\frac{1}{6}$), then one will take this to mean that the assertion is absurd.

In a number of situations we can determine probabilities *a priori* (without undertaking an experiment). Often this can be achieved by symmetry considerations, as in the example of dice above, where the symmetry of the die means that the six is one of 6 possible results which we take to have the same probability. There is an extension of our empirical basis, to the effect that in such situations we get agreement with the frequency interpretation, i.e., it turns out that the relative frequency stabilises precisely around the probability given beforehand, the *a priori probability.*

However, we cannot determine the a priori probability in all situations, and this is where the frequency interpretation is of great significance. It is what gives us a meaningful way of speaking of the probability of decay of an atom of a radioactive isotope within a specified time interval, or of the probability of a given vote going to a particular party, or of a great range of probabilities in technology, science and everyday life in broad generality.

The frequency interpretation also sets natural limits on what we can (or ought to) handle in probability theory. Thus it is (cf. Thesis 11) not immediately obvious that we can give a meaning to the probability of the sun rising tomorrow, or to the probability of a core meltdown at an atomic power plant.[4]

The empirical fact of the stabilisation of relative frequency has an important extension to situations where the experiment is associated with a number (a random variable). The latter could be the sum of the numbers thrown with the die or the waiting time until a radioactive decay occurs. Let us again think of making a long series of trials. In that case we can compute the *empirical average*, i.e., the sum of the numerical values associated with the individual trials, divided by the number of trials. Experience shows that when more and more trials are made, the empirical average stabilises around a certain value, the *theoretical average*, or as we normally prefer to say, *the expectation* or *mean value* or just *mean*. It is this state of affairs that is expressed by the second half of Thesis 12.

When we speak of the stabilisation of the relative frequency and the empirical average as empirical facts, there is a certain vagueness, inasmuch as the stabilisation only occurs "in the limit." This does not mean we have to make infinitely many trials, but strictly speaking the stabilisation laws demand that an arbitrarily large number of trials be undertaken, and this demand is more than we humans can live up to. The straitjacket of finiteness which confines us can only be burst open in the mathematical world.

Thus, strictly speaking, we also cannot uphold the view that the stabilisation laws are purely *empirical* facts. Nevertheless, it would be a completely unreasonable attitude (particularly in the light of history) to reject the stabilisation laws as empirical facts. The laws have crystallised in the world of experience (here we can point to the experience gathered over the course of time with games of chance) even though experience does not always lead to the correct conclusions. And in fact, is it not a characteristic of experience, that it does not give complete

---

[4]To point to a recent example of abuse of probability considerations, we mention the Challenger catastrophy. The reader may consult the illuminating account in the book by Richard P. Feynman: "What Do *You* care What Other People Think?" (W.W. Norton & Company, New York 1988). There it is disclosed that the NASA administration claimed a probability as low as 1 in 100,000 for failure of a mission with Challenger, whereas some NASA scientists had pointed to 1 in 300 as a more realistic estimate. Permit me to digress and state some principles concerning events with low probability: In theory, only events with zero probability do not happen, whereas in normal statistical practise also events with positive, but low, probability do not happen. However, if it is claimed that an event has ultra low probability, be aware! Such events just may happen, as we have learned from painful experience. This observation does not question the validity of probability but merely warns against uncritical reference to this theory in all kinds of situations.

certainty?

We can also defend our view by drawing a parallel with an entirely different field, namely geometry. Here most people feel that it is firmly based in experiences concerning, e.g., the concept of distance. But if we think more deeply about it we see that, here too, experience is unreliable. The notion of exact length cannot, strictly speaking, be regarded as an empirical fact, since it demands more and more precise measurements, which again implies bursting out of the straitjacket of finiteness.

When we compare the young theory of probability with the ancient one of geometry we must admit that both build on experiences of a somewhat insecure kind. But many circumstances make probability theory less accessible. For example, we can point to the fact that while the geometric concept of length concerns concrete material objects, the same is not true of probability. We cannot directly associate a material object with a probability and use our senses to measure the probability in question. The ruler of probability theory – given as it is by the stabilisation laws – is of a more complicated kind than that of geometry.

The above-mentioned circumstances help explain the fact that probability theory is a young science. Another important factor is that in earlier times there was a refusal to study probability seriously, on philosophico-religious grounds. Without going further into this, it may be mentioned that Aristotle explicitly excluded theorising about the concept of chance from the scholarly Parnassus.

As we said earlier, Thesis 12 is expressed in the mathematical world in the form of the law of large numbers, and in the real world by the laws of stabilisation. It must be stressed that the fact that it is possible to prove the law of large numbers in no way shows that the laws of stabilisation are correct. On the contrary, it is more the case that our experience with the stabilisation laws leads us to *demand* a mathematical theory of probability that includes suitable concepts for the formulation of theorems that reflect the stabilisation laws and – naturally – that make it possible to prove these theorems.

After this general discussion of the empirical basis of probability theory, the question remains: what is the content of a stochastic model?

THESIS 13. *The most important constituents of stochastic models are random variables.*

I remind you of the formal definition: a random variable is a real-valued function defined on a sample space.

Perhaps it is surprising that the concept of a probability space does not stand out as the most important. The situation can be thrown into relief by the following.

THESIS 14. *Random variables are given by nature. Probability spaces are fictitious mathematical constructions.*

In the concrete problem we are concerned with, it is fairly clear what this means. The "natural" random variables are the arrival times t_n, the waiting times v_n and the number variables $N(t)$. We are agreed that the phenomenon is nondeterministic and our attempt to set up a stochastic model proceeds simply by regarding these quantities as random variables. We must distinguish clearly between the actual observed values of the t's, v's and N's, on the one hand, and the t's, v's and N's regarded as random variables, on the other.

The problem itself gives us some random variables, but there is nothing in the presentation of the problem to tell us which probability space it is natural to work with; in fact we cannot even set up a sample space. We therefore have the curious situation that the problem immediately gives us some random variables to work with, without giving us domains of definition for them!

The view expressed above can be rejected as rubbish inasmuch as a function can be given only when one knows the domain of definition and the image set. At the same time, if one wants to uphold the close connection between probability theory and reality (cf. Thesis 16 below), then one would be wise to accept the view behind Thesis 14. This is not to say that probability spaces are inessential. Certainly not. Later (§18) we shall come to their most important function, and in any case the following may help to explain.

The domain of definition of a random variable is not of decisive importance. What matters is the *distribution*. The distribution of a random variable X allows us, loosely speaking, to determine the probabilities with which X assumes certain values. More precisely, the distribution is defined to be the map

$$A \mapsto P(X \in A), \tag{3}$$

where A denotes a subset of $\mathbf{R}$. In (3), P refers to the probability function. Here $P(X \in A)$ is an abbreviation for

$$P(\{\omega \in \Omega \mid X(\omega) \in A\}),$$

where Ω denotes the sample space. To be able to write (3) we *still* have to know the probability space – or do we? It turns out that in many situations, in particular the problem we are interested in, the distributions can be determined without explicitly determining the probability space. Let us summarise:

THESIS 15. *A random variable is determined by its distribution.*

In accordance with this we say that two random variables are *identically distributed* when they have the same distribution, and in that case they are considered the same from the point of view of probability theory. Notice also that the definition is meaningful when the random variables are defined on completely different sample spaces.

When X is a *discrete random variable*, i.e., when there is a countable set $A \subseteq \mathbf{R}$ such that $P(X \in A) = 1$, then the distribution is completely determined by the values $P(X = a)$ for $a \in \mathbf{R}$ (all but a countable set of which are 0). Thus if we wish to know the distribution of the $N(t)$'s in our radioactivity problem, we have to determine the probabilities $P(N(t) = k)$ for each value of $t > 0$ and each value $k = 0, 1, 2, \ldots$. Determining these numbers will in fact be our main task.

For an arbitrary random variable X the *distribution function*, defined to be the function

$$x \mapsto P(X \leq x); \qquad x \in \mathbf{R}$$

completely determines the distribution. We shall not prove this. The distribution function is normally denoted by the letter F, thus

$$F(x) = P(X \leq x); \qquad x \in \mathbf{R}.$$

E.g., for the waiting time in our problem the distribution is naturally determined by the distribution function.

Since the distribution function F, by what we have just claimed, contains all probability theoretic information about the random variable X, it must in particular be the case that we can compute the mean value from knowledge of F alone. This is indeed correct. When $X \geq 0$ there is a simple formula for the mean value, namely

$$E(X) = \int_0^\infty P(X > x)dx = \int_0^\infty (1 - F(x))dx. \tag{4}$$

Here E denotes the mean value ("expected value"). Formula (4) applies in particular to all random variables of interest to us (t's, v's and N's). A proof of (4) is sketched in Exercise 11 (however, the proof must be taken with certain reservations for a nondiscrete random variable).

To conclude this section on stochastic models it is appropriate to state:

THESIS 16. *The aim of probability theory is to set up and study models of stochastic phenomena. The main ingredients of such models are random variables.*

4. Time Invariance

We aim to set up a stochastic model which gives answers to questions such as the following:

1) What is the probability that at least 9 events ("clicks") occur before 1 minute has passed?
2) What is the probability of the first and second waiting times being under 1 minute?
3) What is the expected number of events in the time interval $(0, t]$?
4) What are the mean waiting times?

In the usual notation (we use the comma for "logical and") the questions can be formulated as:

1) $P(N(1) \geq 9) = P(t_9 \leq 1) = ?$
2) $P(v_1 \leq 1,\ v_2 \leq 1) = ?$
3) $E(N(t)) = ?$
4) $E(v_1) = ? \quad E(v_2) = ? \quad E(v_3) = ?$

From the outset it might appear to be an impossible task to find the answers to all such questions. One possibility is to take the empirical view (Thesis 12). This has the drawback that each new question demands a new empirical determination. For this reason, among others, we therefore are eager to find a theoretical model. However, we cannot manage entirely without experiments. Later we shall see that we need only one empirical determination, namely that of the intensity. Knowledge of the latter, together with some structural assumptions, will lead to the goal.

In this section and the next, we shall formulate the essential structural properties. Let us take as our starting point the above-mentioned questions.

The most obvious comment concerns 4), where we note that

$$E(v_1) = E(v_2) = \dots , \tag{5}$$

because, according to what we mentioned earlier, the system is unchanged during the observations.

This important property can be illustrated, e.g., as follows. We observe a radioactive preparation and choose a starting time. This will be our time 0. A friend, who really should have joined in the observation, comes late, arriving at the later time T_0 . He will only observe the process from time T_0. Therefore, T_0 is our friend's starting time and will be time 0 according to his reckoning.

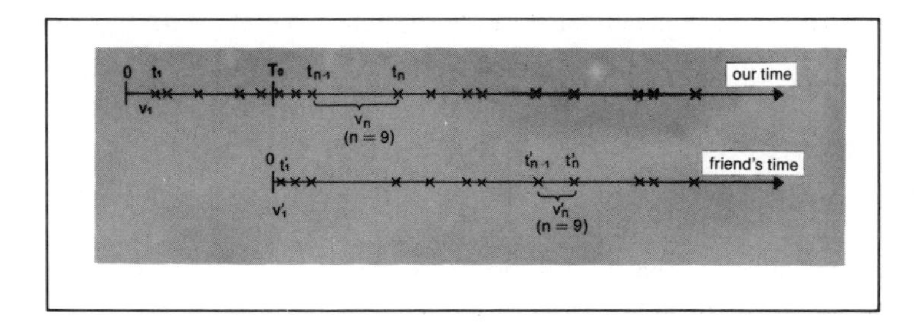

Fig. 2.

The assumption we have in mind can be formulated as follows: the observations are not exactly the same from the two viewpoints – e.g., for the situation in Figure 2 our first waiting time is longer than our friend's – but the two observations behave exactly the same *stochastically.* Herein lies, among other things, the assumption that, although v_1 can be longer than v_1' for particular observations (we use primes to indicate quantities seen by our friend), it will be true for each number s that the probabilities $P(v_1 \leq s)$ and $P(v_1' \leq s)$ are identical. In writing down these two probabilities we are of course not taking v_1 and v_1' to be concrete numbers, but random variables. Thus v_1 and v_1' are identically distributed.

To avoid any misunderstanding, I shall elaborate on what underlies the assumption $P(v_1 \leq s) = P(v_1' \leq s)$, so that it will become clear how the assumption can be verified experimentally. To that end, we imagine that we make observations every day on a radioactive material that is prepared in the same way every day. And every day our friend disappoints us by coming late, so that he always begins observations at a starting time T_0 later than ours. Let us say that we carry out observations over N days, where N is a large number. Our relative frequency corresponding to the event "$v_1 \leq s$" is the number of days on which we do not wait longer than s time units before the first click, divided by N. Our friend's relative frequency corresponding to the event "$v_1' \leq s$" is defined similarly. The assumption amounts to the same thing as saying that the two relative frequencies are equal, to a high approximation.

The assumption we are working towards can be formulated as follows: *the system is, from the stochastic viewpoint, time invariant.*

We shall give yet another formulation of this hypothesis. If we think again of our friend who begins observations late, we can say that if the process is "stopped" at time T_0 and we begin observations anew at that moment (this corresponds to our friend's observations), then the behaviour is the same, from the stochastic viewpoint, as if we had observed continuously. We refer to T_0 as a *stopping time.*

Until now we have had only a *deterministic stopping time* in mind, i.e., T_0 has been given in advance. Now we shall extend the assumption of time invariance by allowing T_0 to be a *stochastic stopping time,* i.e., we shall allow T_0 to depend on the course of the observation. E.g., T_0 can be the moment of the nth event, so $T_0 = t_n$, where n is a fixed natural number.

ASSUMPTION A_1 (TIME INVARIANCE). *If observation is begun anew at a stopping time T, then the observation is the same, from the stochastic viewpoint, as the unstopped observation. And this holds whether T is a deterministic or stochastic stopping time.*[5]

Let us apply A_1 with $T_0 = t_{n-1}$. For an observation that starts at time T_0 we find that the waiting time for the first event equals v_n, and hence equals the waiting time for the nth event in the "unstopped" observation. It follows from A_1 that v_1 and v_n must be identically dis-

[5] It may be interesting, for those who would like to know more exactly what a stopping time is, to mention the essential demand that one never stops at a moment defined by events that occur later. Thus $T_0 = \frac{1}{2}t_1$ is not a stopping time, but $T_0 = t_1$ and $T_o = 2t_1$ are stopping times. More mathematically we demand that T_0 be a random variable and that for each number $s > 0$ the event $\{T_0 > s\}$ depends only on the process before time s; in our case this can be expressed by saying that we can decide whether $T_0 > s$ from knowledge of the random variables $N(t)$ for $t \leq s$ alone.

tributed. *Waiting times are therefore identically distributed.* Formula (5), among other things, follows from this.

As a second consequence of A_1, we notice that *the number of events in intervals of the same length are identically distributed.* Namely, let I be an interval of length t, say $I = (s, s + t]$, and set

$$N(I) = \text{ number of event in } I. \tag{6}$$

Applying A_1 with $T_0 = s$, it follows that $N(t)$ and $N(I)$ are identically distributed. Think it over!

5. Independence

Our knowledge of the physical system we wish to describe leads us to make some more assumptions:

ASSUMPTION A_2 (INDEPENDENCE OF WAITING TIMES). *The waiting times are independent random variables.*

ASSUMPTION A_3 (INDEPENDENCE OF THE NUMBERS OF EVENTS). *If $I_1, I_2, \ldots, I_k$ are non-overlapping intervals on the time axis, then the random variables that give the numbers of events in the different intervals are independent.*

It is appropriate to recall the definition of independence. The latter is first defined for events. The definition is most often given in the following form: two events A and B are *independent* if $P(A \cap B) = P(A) \cdot P(B)$. Three events A, B and C are *independent* if A and B are independent, B and C are independent, C and A are independent and if $P(A \cap B \cap C) = P(A) \cdot P(B) \cdot P(C)$. Corresponding definitions are set up for more events. We can even define independence for a whole sequence of events $A_1, A_2, A_3, \ldots$ by saying that for each choice of finitely many indices $n_1 < n_2 < \cdots < n_k$

$$P(A_{n_1} \cap A_{n_2} \cap \cdots \cap A_{n_k}) = P(A_{n_1})P(A_{n_2}) \cdots P(A_{n_k}).$$

The idea behind these exact definitions is of great importance. It is clearest when we think of two events A and B. The idea is that A is considered to be independent of B when the probability we ascribe to A is not altered by the information that B has occurred. This is expressed by the equation $P(A) = P(A \mid B)$, where $P(A \mid B)$ is the *conditional probability of A given B*. Since $P(A \mid B)$ is given by the intuitively convincing formula

$$P(A \mid B) = \frac{P(A \cap B)}{P(B)},$$

we finally must have (for $P(B) > 0$) that $P(A \cap B) = P(A) \cdot P(B)$. The symmetry of this formula tells us, as was by no means obvious beforehand, that if A is independent of B then likewise B is independent of A. Hence we normally use the neutral terminology, saying that A and B are independent.

And now for random variables! When shall we say that random variables X and Y are independent? Intuitively, this should mean that knowledge of the value of one should not affect our estimate of the value of the other. The exact definition is that X and Y are *independent* when, for each choice of real numbers s and t, the events $\{X \leq s\}$ and $\{Y \leq t\}$ are independent.

This definition extends naturally to any finite or denumerable set of random variables. Thus $X_1, X_2, X_3, \ldots$ are independent if the events $\{X_1 \leq s_1\}$, $\{X_2 \leq s_2\}$, $\{X_3 \leq s_3\}$, are independent for each sequence $s_1, s_2, s_3, \ldots$ of real numbers.[6]

It follows from A_2 that the waiting times $v_1, v_2, \cdots$ are independent and this, combined with A_1, shows that $v_1, v_2, \cdots$ *is a sequence of independent identically distributed random variables.*

---

[6]One can show that this implies that, for any sequence $B_1, B_2, \ldots$ of subsets of $\mathbf{R}$, the events $\{X_1 \in B_1\}, \{X_2 \in B_2\}, \ldots$ are independent.

6. Intermezzo on the Binomial Distribution

In connection with the properties found for the waiting times, it is worth mentioning that a sequence of identically distributed independent random variables $X_1, X_2, \ldots$, or perhaps just a finite sequence $X_1, X_2, \ldots, X_n$, is very often the natural starting point for investigations in probability theory.

A well-known example of this is where the X's are *success variables,* also called *Bernoulli variables* or *Bernoulli trials*; here it is understood that the X's can take only the values 0 and 1. In such a case the interpretation of the event $\{X_k = 1\}$ is "success at the kth trial," e.g., "head on the kth throw of a coin," "six on the kth throw of a die," "girl at the kth attempt," "defective nut on inspection of the kth nut by spot test," or the like.

The assumption that the X's are identically distributed implies that the probability $P(X_k = 1)$ of success at the kth trial does not depend on k; this probability is called the *success probability.* If the success probability is p, then $q = 1 - p$ is the probability of "failure" in a single trial. For a success variable X with success probability p we find the mean to be

$$E(X) = 0 \cdot P(X = 0) + 1 \cdot P(X = 1) = p. \tag{7}$$

The distribution of X is called the *Bernoulli distribution with parameter* p.

The assumption of independence allows all possible probabilities to be computed, e.g.,

$$\begin{aligned} &P \text{ (at least one success in the first 3 trials)} \\ &\quad = 1 - P(\text{failure at the first 3 trials}) \\ &\quad = 1 - P(X_1 = 0, X_2 = 0, X_3 = 0) \\ &\quad = 1 - P(X_1 = 0) \cdot P(X_2 = 0) \cdot P(X_3 = 0) \\ &\quad = 1 - q^3. \end{aligned}$$

The binomial distribution with parameters n and p is defined to be the distribution of the number of successes in n trials, where the success probability of each trial is p; it is also required that the trials in question be independent. More mathematically then, we can say that the binomial distribution with parameters (n, p), where $n \in \mathbf{N}$ and $0 < p < 1$, is the distribution of the random variable

$$S = X_1 + X_2 + \cdots + X_n, \tag{8}$$

where $X_1, X_2, ..., X_n$ are independent success variables, each with success probability p.[7]

Thus, the binomial distribution with parameters $(1, p)$ is the same as the Bernoulli distribution with parameter p.

The binomially distributed random variable S in (8) can assume the values $0, 1, 2, \ldots, n$, and the distribution is completely specified by giving the probability that $S = k$ for each possible value k.

The *binomial coefficient*

$$\binom{n}{k} = \frac{n!}{k!(n-k)!} = \frac{1}{k!}n(n-1)\cdots(n-k+1) \tag{9}$$

gives the number of solutions of the equation $x_1 + x_2 + \cdots + x_n = k$, where the x's can take the values 0 or 1. Each of these solutions corresponds to a contribution of $p^k(1-p)^{n-k}$ to the probability $P(S = k)$; e.g., the solution $x_1 = x_2 = \cdots = x_k = 1$, $x_{k+1} = x_{k+2} = \cdots = x_n = 0$ contributes

$$\begin{aligned} &P(X_1 = 1, X_2 = 1, \ldots, X_k = 1, X_{k+1} = 0, \ldots, X_n = 0) \\ &\quad = P(X_1 = 1)P(X_2 = 1) \cdots P(X_k = 1)P(X_{k+1} = 0) \cdots P(X_n = 0) \\ &\quad = p^k \cdot (1-p)^{n-k}. \end{aligned}$$

---

[7]In certain accounts one sees a definition of the binomial distribution for which knowledge of the probability space is also required. This is highly unfortunate and in conflict with the attitude formulated above, in Thesis 15, for example.

In this way we find the well-known formula for the binomial distribution:

$$P(S = k) = \binom{n}{k} p^k (1-p)^{n-k}; \qquad k = 0, 1, \ldots, n. \tag{10}$$

Before we leave the binomial distribution, let us recall how to compute its mean. Since S can assume only the values $0, 1, 2, \ldots, n$, the definition says that

$$E(S) = \sum_{k=0}^{n} k \cdot P(S = k),$$

which in principle can be worked out with the help of (10). However, it is much easier to use the fact that the mean of a sum is the sum of the means — an intuitively obvious property, though not a direct consequence of the definition (see Exercise 9). This property, in combination with (7), directly gives

$$\begin{aligned} E(S) &= E(X_1) + E(X_2) + \cdots + E(X_n) \\ &= n \cdot E(X_1) \\ &= np. \end{aligned} \tag{11}$$

This intermezzo may appear a little unmotivated. What has the binomial distribution to do with the problem at hand? That is what we are going to see!

7. On the Physical Background to the Assumptions

We have previously called on knowledge of physics in order to motivate our investigations. Nevertheless, we shall add a few remarks here; they will not be seen as completely satisfying from the physical standpoint – if anything they are "pseudophysics."

We suppose that the radioactive preparation we study consists of a large number of atoms and that only a vanishingly small fraction of these decay under observation. In addition, we assume that each individual atom can be found in one of two states, the excited state of high energy, and the ground state of low energy. An atomic nucleus can decay from the excited state to the ground state and each individual decay registers on our counter, e.g., as a click.

The assumptions are an idealisation. Normally, radioactivity is due to several processes, e.g., there can be several excited states and so "cascade decay" can occur, or proper nuclear transformation can occur (e.g., with α- or β-active sources).

The history of an individual atomic nucleus is shown schematically in Figure 3, with t_0 denoting the moment of decay.

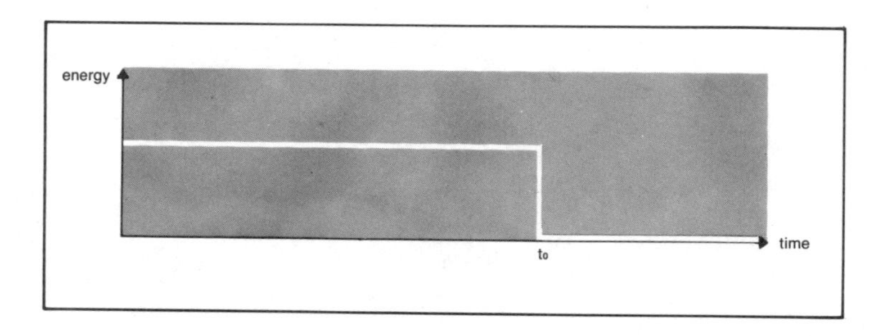

Fig. 3.

The most conspicuous feature of the phenomenon is perhaps the *spontaneous character* of the decay. Let me elaborate. If an atomic nucleus has reached the ground state, then it remains there. There is no "motion" towards the moment of decay, since there is no change of state in the individual nucleus in the time interval from 0 to t_0. There is no "ageing effect," the nucleus does not decay to the ground state out of "tiredness." Another way to put it is that no "ripening" process occurs in the time interval from 0 to t_0. Thus the decay occurs as a "spontaneous impulse."

The above refers to individual atomic nuclei. Another important point is that, at least as far as we can tell, the nuclei act independently of each other. When one nucleus decays it makes no "impression" on the remainder and does not affect their tendency to decay.

If one goes through assumptions A_1, A_2 and A_3 again, one will see their reasonableness for the situation just described. It should be mentioned that the above considerations also are central in other approaches to the goal of finding a mathematical model for radioactivity (see Exercise 16).

The main text of this book treats only the case where the intensity is, or can be reckoned to be, constant in time. In applications, as well as in many interesting theoretical investigations, it is important to be able to handle situations where this condition is not satisfied. This is done in the exercises (see especially Exercise 38, where the starting point is the above-mentioned situation regarding spontaneity and independence,but with a new quantity (half-life) also coming into play).

8. The Intensity

The expected number of events per unit time is called the *intensity* and is denoted by λ (Greek "lambda"). This is the only numerical quantity we need to know in order to set up the desired model, building on assumptions A_1, A_2 and A_3 .

Let us first convince ourselves that we have understood the definition. Since $N(1)$ gives the number of events in the time interval $(0, 1]$,

$$\lambda = E(N(1)).$$

The intensity will be determined experimentally. Here it is obvious how to proceed. We observe the process over time intervals of length 1 and, for each time interval, note the number of events. In other words, we observe the concrete values of the following random variables (the first of which is just $N(1)$):

$$N((0,1]), \quad N((1,2]), \quad N((2,3]), \quad \cdots, \quad N((n-1,n]).$$

Since these are independent and identically distributed by A_1 and A_3, the empirical frequency for large n will be a good approximation to the theoretical mean, and hence to the intensity.

If it is more convenient to observe the process over time intervals of length other than 1, it is still possible to determine the intensity. To see this, note that

$$E(N(I)) = s\lambda \tag{12}$$

for each time interval I of length s. This formula merely says that the expected number of events in an interval is proportional to the length of the interval (see Exercise 21).

The intensity can also appear in a slightly different way, namely with the determination of the *mean waiting time* μ (Greek "mu"), which is defined by

$$\mu = E(v_1) = E(v_2) = \dots . \tag{13}$$

Since $v_1, v_2, \dots$ are independent and identically distributed, it must in principle be possible to determine μ from the actual observed value of

$$\frac{v_1 + v_2 + \cdots + v_k}{k}, \tag{14}$$

where k is a large number. Here we are relying on the law of stabilisation of the empirical average, cf. the discussion in connection with Thesis 12.

We now remark that

$$\lambda = \frac{1}{\mu}, \tag{15}$$

because if we have observed values $v_1, v_2, \dots, v_k$ it follows that there are k events in $v_1 + v_2 + \cdots + v_k$ units of time, hence an average number of $k/(v_1 + \cdots + v_k)$ events per unit time. But this number is just the reciprocal of the number in (14). See also Exercise 11 (or Exercise 13) for a more theoretical derivation of the formula $\mu = 1/\lambda$ for the mean waiting time.

9. Mathematics, at last!

The main result we are aiming at is the determination, for each value $t > 0$, of the distribution of $N(t)$, i.e., we have to determine the numbers

$$P(N(t) = k); \qquad k = 0, 1, 2, \dots . \tag{16}$$

We have been on the way for a long time and a surprise is still in store, in the form of an extra assumption, before we arrive at the result.

In what follows we fix the value of t.

The structure we have agreed on with assumptions A_1, A_2 and A_3 must be expressed in the strongest possible way when we try to determine the probabilities in (16). We shall do this by dividing the interval $(0, t]$ into subintervals and making use of the fact that $N(t)$ is the sum of the numbers of events in the subintervals.

We choose an $n \in \mathbf{N}$ and divide $(0, t]$ into n equal intervals

$$I_1 = \left(0, \tfrac{t}{n}\right], \quad I_2 = \left(\tfrac{t}{n}, \tfrac{2t}{n}\right], \dots , \quad I_n = \left(\tfrac{(n-1)t}{n}, t\right].$$

Thus

$$N(t) = N(I_1) + N(I_2) + \dots + N(I_n). \tag{17}$$

By our assumptions, $N(I_1)$, $N(I_2)$, $\dots$, $N(I_n)$ are independent and identically distributed. Since they are identically distributed we shall have, for each $k = 0, 1, 2, \dots$,

$$P(N(I_1) = k) = P(N(I_2) = k) = \dots = P(N(I_n) = k). \tag{18}$$

From (12) we also have

$$E(N(I_1)) = E(N(I_2)) = \cdots = E(N(I_n)) = \frac{\lambda t}{n}. \tag{19}$$

So far we have not obtained anything, since $N(t)$ remains written as a sum of independent random variables of exactly the same type as $N(t)$ itself – and that is hardly a simplification, rather the contrary.

However, there is one difference, namely that the random variables $N(I_1)$, $N(I_2)$, $\ldots, N(I_n)$ give the numbers of events in smaller intervals than that for $N(t)$ and, by choosing n large, we can arrange for the lengths of the intervals I_1, I_2, $\ldots, I_n$ to be as small as we please.

This freedom to choose n as we please – formula (17) being exact for each n – can be used to choose n so large that t/n, the length of the intervals $I_1, \ldots, I_n$, is much less than the mean waiting time μ. Then our common sense tells us that there is at most one event in each of the subintervals $I_1, \ldots, I_n$.

If we therefore choose n to be sufficiently large we can reckon that $N(I_1)$, $N(I_2)$, $\ldots, N(I_n)$ take only the values 0 and 1. This cannot be absolutely correct; there will always be a positive, but very small probability of two (or more) events in one of the intervals I_1, I_2, $\ldots, I_n$. But it must be possible to replace the random variables $N(I_1), \ldots, N(I_n)$ by new random variables $N^*(I_1), \ldots, N^*(I_n)$, each of which can take only the values 0 and 1, and obtain that the random variable defined by

$$N^*(t) = N^*(I_1) + N^*(I_2) + \cdots + N^*(I_n) \tag{20}$$

has a distribution closely approximating that of $N(t)$.

If we are to give a precise meaning to these considerations, we must make some assumption about $N^*(I_1), \ldots, N^*(I_n)$ beyond the one already made that they are success variables. Since $N(I_1), \ldots, N(I_n)$ are independent, it is natural to ask that $N^*(I_1), \ldots, N^*(I_n)$ be independent as well. And since $N(I_1), \ldots, N(I_n)$ all have mean value $\lambda t/n$, it is natural to ask the same of $N^*(I_1), \ldots, N^*(I_n)$, i.e., the success probability of these random variables is stipulated to be (cf. (7))

$$p = \frac{\lambda t}{n}. \tag{21}$$

In short, we let $N^*(I_1), \ldots, N^*(I_n)$ be independent success variables with success probability given by (21) and claim that the distribution of $N^*(t)$, defined by (20), is a good approximation to that of $N(t)$ for n large.[8]

But we know $N^*(t)$'s distribution! This is the binomial distribution with parameters n and p, hence it is (see (10))

$$(22)\quad P(N^*(t) = k) = \binom{n}{k}\left(\frac{\lambda t}{n}\right)^k \cdot \left(1 - \frac{\lambda t}{n}\right)^{n-k}; \qquad k = 0, 1, 2, \ldots, n.$$

In the above we have had a fixed n in mind, hence the dependence of $N^*(t)$ on n is not expressed notationally in (22). But $N^*(t)$ obviously does depend on n, and this can also be seen from the right-hand side of (22).

The discussion has convinced us that by passing to the limit as $n \to \infty$ in (22) we must obtain the desired probability. Furthermore, we have argued for an assumption that secures this, namely the following:

Assumption A_4 (a regularity condition). *For each $t > 0$,*

$$\lim_{n\to\infty} P\left(\bigcup_{k=1}^{n} \{N((\tfrac{(k-1)t}{n}, \tfrac{kt}{n}]) \geq 2\}\right) = 0.$$

By going to the complementary event, the assumption can also be formulated as

$$(23)\qquad P\left(\bigcap_{k=1}^{n} \{N((\tfrac{(k-1)t}{n}, \tfrac{kt}{n}]) \leq 1\}\right) \to 1 \quad \text{as } n \to \infty.$$

Assumption A_4 appears to be harmless. If we explain to a physicist what the assumption means and ask him whether we may make this assumption, there will be no doubt in his mind. Of course!

But why have we not mentioned this assumption earlier, with the other assumptions? The reason is that no one would have dreamt of presenting such an assumption in advance. This is clearly an assumption that first emerges from the mathematical analysis. It is harmless

[8]Note that we have not made any demands on the probability space associated with $N^*(I_1), \ldots, N^*(I_n)$. This is because it is only the *distribution* which is sought. If one *insists* on working with the same probability space that is associated with $N(I_1), \ldots, N(I_n)$, then one can set $N^*(I_k) = 0$ when $N(I_k) = 0$ and $N^*(I_k) = 1$ when $N(I_k) \geq 1$ (however, this makes the success probability a little different from that given in (21)).

from the physicist's standpoint, and is merely to ensure that the mathematical machinery runs smoothly. One often refers to such assumptions as *regularity conditions.*

Regularity conditions normally interest only mathematicians; for physicists (the users) it does not matter what assumptions of this kind we make. In our case we could readily assume, e.g., that the distribution function for the waiting time, $x \mapsto P(v_1 < x)$, is differentiable arbitrarily often for $x > 0$.

THESIS 17. *Regularity conditions are pedantries from the user's viewpoint, but necessary for the mathematical analysis and understanding.*

There is an important reason for taking careful account not only of natural structural assumptions, but also of regularity conditions. Namely, if there comes a day when the user is shocked by a phenomenon that is expected to follow a certain mathematical model, while observations show that it does not, then it *may* happen to be just because a regularity condition does not hold, and perhaps the possibility of "irregular" behaviour in the mathematical respect will reveal new phenomena of purely physical – or other practical – interest.

Having said that, it must be admitted that, unfortunately, the applied mathematician often has to curb "pedantry" and thus renounce an analysis that is complete in all mathematical detail. There may simply be too many profound theoretical aspects to deal with – if you want to arrive at a result, that is. The applied mathematician must therefore have great theoretical insight and experience to be able to separate out subordinate theoretical problems and devote time only to problems that are important for the concrete application in question.

We now lay philosophy aside and proceed to the limit as $n \mapsto \infty$ in (22). First we write (22) in the form

$$
\begin{aligned}
&P(N_n^*(t) = k) \\
(24)\quad &= \frac{(\lambda t)^k}{k!} \cdot \left(1 - \frac{1}{n}\right)\left(1 - \frac{2}{n}\right) \cdots \left(1 - \frac{k-1}{n}\right)\left(1 - \frac{\lambda t}{n}\right)^{-k}\left(1 - \frac{\lambda t}{n}\right)^{n}.
\end{aligned}
$$

Here we have written $N_n^*(t)$ in place of $N^*(t)$ because the dependence on n is now essential. We point out that t continues to be fixed. It is also important that we are considering a fixed k. Evidently

$$\lim_{n\to\infty}\left[\frac{(\lambda t)^k}{k!}\cdot\left(1-\frac{1}{n}\right)\left(1-\frac{2}{n}\right)\cdots\left(1-\frac{k-1}{n}\right)\cdot\left(1-\frac{\lambda t}{n}\right)^{-k}\right] = \frac{(\lambda t)^k}{k!}. \tag{25}$$

It remains merely to investigate the behaviour of $\left(1-\frac{\lambda t}{n}\right)^n$ as $n\to\infty$. This is easy as soon as one has the idea of looking at logarithms. One then get:

$$\lim_{n\to\infty}\left(1+\frac{a}{n}\right)^n = e^a \qquad \text{for all real } a. \tag{26}$$

PROOF OF (26): By the continuity of the exponential function it is enough to show that

$$\lim_{n\to\infty}\left[n\cdot\log\left(1+\frac{a}{n}\right)\right] = a.$$

But the latter follows immediately by writing

$$n\cdot\log\left(1+\frac{a}{n}\right) = a\cdot\frac{\log\left(1+\frac{a}{n}\right)-\log 1}{\frac{a}{n}}$$

and from the knowledge of the derivative at $x=0$ of the function

$$x\mapsto\log(1+x). \qquad \blacksquare$$

From (24), (25), (26) and the following limiting equation, which was discussed so thoroughly above,

$$\lim_{n\to\infty}P(N_n^*(t)=k) = P(N(t)=k),$$

we get

THEOREM 1. *Under assumptions A_1, A_2, A_3 and A_4 we have the following, where λ denotes intensity:*

(i) *For each $t > 0$ and $k = 0, 1, 2, \ldots,$*

$$P(N(t) = k) = \frac{(\lambda t)^k}{k!} e^{-\lambda t}.$$

(ii) *For $0 < s < t$ the random variables $N(t) - N(s)$ and $N(t - s)$ are identically distributed.*

(iii) *For $0 < s_1 < s_2 < \cdots$ the random variables*

$$N(s_1), \quad N(s_2) - N(s_1), \quad N(s_3) - N(s_2), \ldots$$

are independent.

In (ii) and (iii) we have reiterated some properties discussed earlier.

The random variables $N(t)$; $t > 0$, make up a whole family of random variables. One says that $(N(t))_{t>0}$ is a *stochastic process.* When the conditions (i), (ii) and (iii) of Theorem 1 hold, $(N(t))_{t>0}$ is said to be a *Poisson process with intensity* λ.

An individual random variable X is said to have a *Poisson distribution* if X can take only integer values $0, 1, 2, \ldots$ and there is a number λ such that

$$P(X = k) = \frac{\lambda^k}{k!} e^{-\lambda}; \qquad k = 0, 1, 2, \ldots. \tag{27}$$

The distribution given by (27) is called the *Poisson distribution with parameter* λ. Evidently $\lambda = E(X)$.

In a Poisson process with intensity λ each individual random variable has a Poisson distribution with parameter proportional to the "time" t, and the proportionality factor is just λ.

Properties (i), (ii) and (iii) in Theorem 1 allow us to work out all probabilities we believe to be interesting. Thus, for the arrival times $t_1, t_2, \ldots$ (which indeed are random variables) and waiting times $v_1, v_2, \ldots$ (likewise random variables) we have the following results:

THEOREM 2. *(i) For each $n = 1, 2, \ldots$ the distribution function of the nth arrival time is given by*

$$P(t_n \le t) = e^{-\lambda t} \sum_{k=n}^{\infty} \frac{(\lambda t)^k}{k!} = 1 - e^{-\lambda t} \sum_{k=0}^{n-1} \frac{(\lambda t)^k}{k!}; \qquad t > 0.$$

(ii) The waiting times $v_1, v_2, \cdots$ are independent and identically distributed, with distribution function given by

$$P(v_1 \leq x) = 1 - e^{-\lambda x}; \qquad x \geq 0.$$

The mean waiting time is

$$E(v_1) = 1/\lambda.$$

PROOF: It is understood, of course, that the conditions of Theorem 1 are satisfied.

(i): This follows from Theorem 1 since

$$P(t_n \leq t) = P(N(t) \geq n) = 1 - P(N(t) < n).$$

(ii): This is partly a repetition of the property we saw earlier, partly a trivial consequence of (i) when $v_1 = t_1$. ∎

The waiting time distribution that appears in Theorem 2 is called the *exponential distribution with parameter* λ. A positive random variable X is said to be *exponentially distributed* when there is a positive number λ such that

$$P(X \leq x) = 1 - e^{-\lambda x}; \qquad x \geq 0. \tag{28}$$

We see that a Poisson process with intensity λ can be characterised as one that gives the number of events up to time t, $t > 0$, where events occur in such a way that the waiting times are independent exponentially distributed random variables with parameter λ. Think it over!

10. Confrontation with Reality

The model we have set up gives good agreement with observed radioactive processes. It would be most convincing to make your own experiments and see whether they agree with theory.

As previously mentioned, it can be difficult in practice – often quite impossible – to carry out precise observations of arrival times. It is easier to choose a suitable length of time, and to observe the numbers of events in non-overlapping intervals of the chosen length. A convenient way to present the results of an observation is to enter the numbers N_k; $k = 0, 1, 2, \ldots$ in a table, where

$$N_k = \text{ number of time intervals containing } k \text{ events.} \tag{29}$$

Let U and V be the quantities

$$U = \text{ number of time intervals observed,} \tag{30}$$

$$V = \text{ total number of events observed.} \tag{31}$$

Thus

$$U = N_0 + N_1 + N_2 + \cdots , \tag{32}$$

$$V = N_1 + 2N_2 + 3N_3 + \cdots . \tag{33}$$

The *observed intensity* λ_{obs}, measured by the number of events per time interval, is therefore given by

$$(34) \qquad \lambda_{\text{obs}} = V/U.$$

In many situations one will prefer to see the number of events per unit time (e.g., per second). In this case the quantity in (34) must be divided by T, the length of a time interval. Our model naturally leads us to expect that the number of events in each time interval will follow a Poisson distribution with parameter λ_{obs}, i.e., with the observed intensity in place of the theoretical (not assumed to be known).

Thus we can compute the probability that a time interval contains k events; multiplying this probability by the total number of intervals observed, we find the expected number of intervals with k events. We call the latter quantity N_k^*. We have

$$(35) \qquad N_k^* = U \cdot \frac{\lambda_{\text{obs}}^k}{k!} e^{-\lambda_{\text{obs}}}; \qquad k = 0, 1, 2, \ldots .$$

Here we are appealing to the law of stabilisation of the relative frequencies. If there is good agreement between the expected number N_k^* and the observed number N_k , we may take this as confirmation that the mathematical model describes the phenomenon satisfactorily. A more cautious conclusion would be that the foregoing investigation does not give cause to reject the conjecture that the phenomenon in question is described by the mathematical model we have set up. In §§16 and 17 we shall comment on a more precise statistical analysis.

I shall not quite leave the reader to make his own observations, but instead mention the results of a famous experiment carried out by Rutherford and Geiger in 1910. They observed a radioactive preparation over $U = 2608$ time intervals, each of length 7.5 seconds. In all, they observed $V = 10097$ events, which gives $\lambda_{\text{obs}} = 3.87$ (number of events per 7.5 seconds). Table 1 shows the results, with the observed numbers (N_k's) against the expected numbers (N_k^*'s) computed from (35) and rounded off to the nearest whole number. The table was worked out by using programs $P1$ and $P2$.

Table 1

k	0	1	2	3	4	5	6	7	8	9	10	11	12	13	14	≥15
N_k	57	203	383	525	532	408	273	139	45	27	10	4	0	1	1	0
N_k^*	54	210	407	525	508	394	254	141	68	29	11	4	1	0	0	0

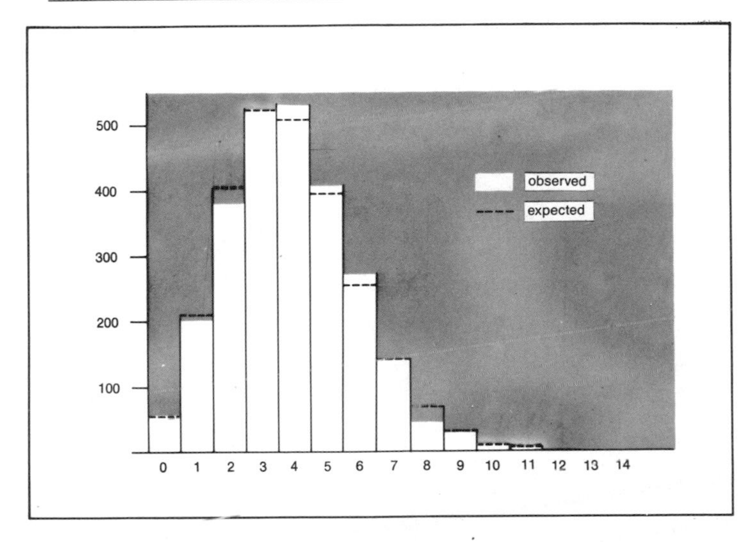

Fig. 4. Histogram for the Rutherford-Geiger experiment.

One can also illustrate the observed and expected values more visually by constructing the associated *histogram*. Figure 4 shows the results.

The most natural reaction to these numbers seems to me to be sheer amazement at the fine agreement between theory and practice. There is even good reason to put it more strongly. If we put ourselves in Rutherford and Geiger's place and think of the amount of work beforehand in organising and carrying out the experiment, then a fair amount of enthusiasm would be in order. On the present basis, we can consider the model to be confirmed. For a more precise statistical analysis we refer to §§16 and 17.

11. Critique of the Mathematics

In reality, our derivation of the main results (Theorems 1 and 2) can be cut to pieces!

The central point of criticism is that we have "forgotten to check the answer." When we solve equations we know that after we have computed and arrived at a (provisional) result, then we have to test and discard possible false roots. The same is true, of course, when we construct models.

Specifically, we must ask whether the Poisson process is a "false root" and therefore to be discarded. That would be extraordinarily annoying, since the Poisson process is the only possibility. A physicist will perhaps conclude, just because it is the only possibility, that there is no reason to make a test, that it must simply be true. We can indeed see that there is a model; we can just watch radioactive decay. However, this argument is not satisfying to a mathematician. True, there is a physical phenomenon to cling to, but what makes us sure it can be described by a mathematical model? We must ask whether probability theory is so wonderful that it allows every stochastic phenomenon in the world to be described mathematically.

It is therefore extremely reasonable to investigate whether the model satisfies the demands we have imposed. In such an investigation it is not the physics which must be tested, but the mathematics. Given that radioactive processes exist, the question is whether we mathematicians can find models for them or not. We cannot dispute the physical phenomenon itself. That is a fact. But until we have constructed the model in all detail and checked that it satisfies our purely mathematical demands, we shall not have made good our mathematical claims.

It is not enough to point out, as we did in the previous section, that the model gives a good description of radioactive processes. Even if we find the model unsuitable to describe reality, it remains a meaningful task to test the model's internal mathematical consistency.

Very well, what should we test? The first thing that springs to mind is perhaps the possibility that the probabilities we found from Theorem 1 are not probabilities at all! Certainly, the values we have given for $P(N(t) = k)$ are positive for all $k = 0, 1, 2, \dots$, but we have not made sure that the sum of these values is 1. This worry may seem exaggerated, since in principle we found the numbers $P(N(t) = k)$ by writing down a schemata of probability vectors:

	column no. k
	$\vec{p}_1 = (p_{10}, p_{11})$
	$\vec{p}_2 = (p_{20}, p_{21}, p_{22})$
	$\vdots$
	$\vec{p}_k = (p_{k0}, p_{k1}, p_{k2}, \dots, \quad p_{kk})$
	$\vdots$
row no. n	$\vec{p}_n = (p_{n0}, p_{n1}, p_{n2}, \dots, \quad p_{nk} \quad , \dots, p_{nn})$
	$\vdots$

and for each k passing to the limit as $n \to \infty$ (set $p_{nk} = P(N_n^*(t) = k)$). Thus it is a question of passing to the limit in the columns (see schemata). However, this in itself does not guarantee that we get a probability vector in the limit; consider, e.g., the example $\vec{p}_n = (0, 0, \dots, 1)$; $n = 1, 2, \dots$. Consequently, there are grounds for a closer look. However, the matter is easily settled by using the formula

$$e^x = \sum_{k=0}^{\infty} \frac{x^k}{k!}, \tag{36}$$

valid for all $x \in \mathbf{R}$ (readers not familiar with this formula can see a proof in Exercise 20).

After this one could go on to investigate whether the model satisfies the demands we imposed in $A_1 - A_4$. However, we shall not go into that here (as far as the regularity condition A_4 is concerned, see Exercise 22).

On the other hand, we point out that if we take a strict view of Theorem 1, it does not actually give a probability theoretic model! For where is the event space? And where is the probability measure? There is no mention of it. And with good reason. We have not thought about it at all!

What we have proved is essentially the following: *if* there is a probability space (Ω, P), and *if* for each $t > 0$ there is a random variable $N(t)$ on Ω which can assume only the values $0, 1, 2, \ldots$ and for which $A_1 - A_4$ hold, then the conditions in Theorem 1 (and Theorem 2) must hold.

There are three remarks to be made about this. First: one *can* find a stochastic model with a probability space, random variables and everything associated with them. Second: this is incredibly difficult. And third: only a pure mathematician would find it worth the trouble!

In the case of the third remark, one should bear in mind the attitude expressed by Theses 13, 14 and 15. Nevertheless, in consideration of the curious person who is eager for knowledge – and one should certainly not disappoint him or her – I shall indicate how the probability space can be constructed.

In choosing the sample space Ω we stick to the idea that has proved fruitful in elementary probability theory, which deals with models of dice throwing and suchlike phenomena. The idea is to let each sample point $\omega \in \Omega$ represent a possible "state," "history," "destiny" or "evolution" – one can choose whichever word one prefers – of the phenomenon in question. This idea makes it natural, in our case, to let Ω be the set of all functions $\omega : \mathbf{R}_+ \to \{0, 1, 2, \ldots\}$ that are nondecreasing and that increase only by one at points where the value jumps.

With this choice of Ω, and $t > 0$, one can define the random variable $N(t)$ to be the mapping $\omega \mapsto \omega(t)$; $\omega \in \Omega$.

It is difficult to define the probability function P so that $N(t)$; $t > 0$ is a Poisson process with intensity λ. However, we shall say a little about it. The idea is to determine $P(A)$ for more and more complicated events $A \subseteq \Omega$. One can obviously begin with $P(\Omega) = 1$ and $P(\emptyset) = 0$. For $t_1 > 0$ and $k_1 \in \{0, 1, 2, \ldots\}$ one can consider the subset

$$\{\omega \in \Omega \mid \omega(t_1) = k_1\} = \{\omega \in \Omega \mid N(t_1) = k_1\}$$

and set

$$P(\{\omega \in \Omega \mid \omega(t_1) = k_1\}) = \frac{(\lambda t_1)^{k_1}}{k_1!} e^{-\lambda t_1}.$$

Next, for $t_1 > 0$, $t_2 > 0$, $k_1 \in \{0, 1, 2, \dots\}$ and $k_2 \in \{0, 1, 2, \dots\}$ one can consider

$$\{\omega \in \Omega \mid \omega(t_1) = k_1,\ \omega(t_2) = k_2\},$$

and determine the probability of the latter event. Figure 5 shows a typical sample point that belongs to this event (the event consists of all ω's that pass through the two crosses).

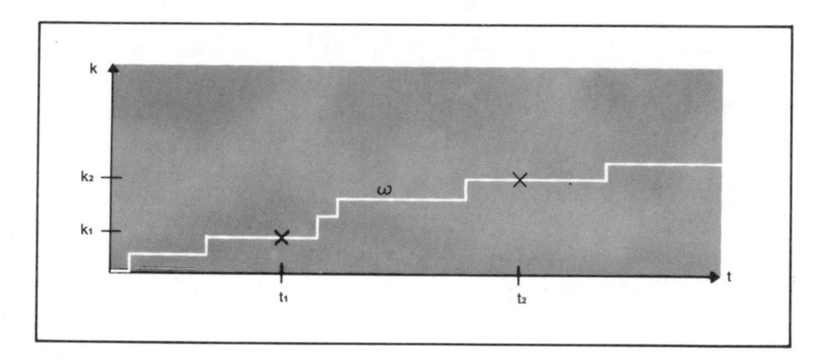

Fig. 5.

For the determination of the probability of this event, see Exercise 29.

In principle one can continue in this way until "eventually" the probability function is determined completely. It is not entirely without complications, but I hope that the above will serve as an indication of the purely mathematical problems that have to be solved in order to set up models of stochastic processes such as the Poisson process. (In §17 we shall return to the formulation of this problem, but in a different and factually simpler situation.)

In this section we have criticised the mathematics, rejecting some specific points of criticism, but admitting others.

An attitude, or rather an acknowledgement, that can be extracted from our discussion is the following:

THESIS 18. *Mathematics is not absolute. Mathematical structures and models can be understood and discussed at several levels.*

12. Critique of the Physics

It would hardly be fair, after having criticised the mathematics so strongly, to let the physics go scot-free. There are two main areas where it is possible to make criticism. The first, and primary one, concerns the phenomenon itself which we are studying. The second, and secondary one, concerns the method of observation. The critique in the secondary area asks whether the apparatus used really allows us to make precise observations of the phenomenon we have decided to study – perhaps there are sources of error that make the observed phenomenon entirely different from the one intended.

Of the primary points of criticism we shall mention three. We first remark that in many radioactive preparations there are several processes, so that the total effect is in fact a "sum," or *superposition* as one says, of several individual processes. Let us suppose that in reality two processes are taking place. One process is described by a Poisson process $(N_1(t))_{t>0}$ with intensity λ_1 and the other by a Poisson process $(N_2(t))_{t>0}$ with intensity λ_2. The total number of events arising from both processes, in the time interval $(0, t]$, is denoted by $N(t)$ and is given by

$$N(t) = N_1(t) + N_2(t); \qquad t > 0.$$

Under the assumption that the two processes do not influence each other, which can be expressed mathematically by saying that "type 1" random variables are independent of the "type 2" random variables, one can conclude that $(N(t))_{t>0}$ is again a Poisson process and that its intensity λ is given by

$$\lambda = \lambda_1 + \lambda_2.$$

This is in fact very easy (see Exercise 25).

We say that $(N(t))_{t>0}$ is the *superposition* of the independent Poisson processes $(N_1(t))_{t>0}$ and $(N_2(t))_{t>0}$.

With this discussion we have realised the following: any Poisson process can in principle arise from the superposition of two or more independent Poisson processes. Nothing in the mathematical model allows us to say anything about the number of constituent "simple processes." Thus it is the physicist's job to say something about this. Even if the physicist has realised that there are precisely two (independent) simple processes, then it can be seen that it is mathematically impossible, from observations of the superposition $N(t) = N_1(t) + N_2(t)$, to say what the two subprocesses contribute individually – as far as their intensities are concerned, one can only say what their sum is.

If one wants to clarify the simple processes in the case just mentioned, then further analysis of the physical circumstances will be required. E.g., it may be that differences in mass, charge or energy conditions will allow the physicist to propose an experimental setup that leads to a separation of the two simple processes, or that suppresses one process by means of a suitable filter. This is just what often happens; e.g. with a source that is both α-and β-active one can normally filter out the α-particles.

The tendency of these observations is to harmonise well with Thesis 3 and the discussion thereafter.

One may also have the situation where there are two, or more, simple processes, that however are dependent. This occurs when the atomic nucleus decays to a stable nucleus via several "intermediate states" as, e.g., in the so-called *radioactive series.*

As a third primary point of criticism we mention that if the half-life is short in comparison with the length of observations, then the intensity cannot be considered constant. A mathematical model that takes account of this is sketched in Exercise 38. Perhaps readers will be interested to hear that the half-life of Polonium 210, which Rutherford and Geiger used in their experiment, is only 138.4 days, and as their observations took place over 5 days, it was necessary to move the source closer to the counter every day to ensure a constant intensity.

The objections just cited are in a way not serious points of criticism. In deriving the main results we have in fact been careful to exclude complications of the kind cited. Therefore the critique does not put a question mark against the model we have set up, but rather against its relevance. Besides, we learn that for deeper investigation there may well be a need for more refined and complex models.

As for points of criticisms of the second kind, we also mention three. The first is based on the fact that most counters cannot eliminate the natural *background radiation.* This means that one is observing the superposition of the proper process, originating from the radioactive preparation, and the background radiation. Since these two processes are clearly independent, and since the background radiation closely follows the Poisson process model, the above-mentioned situation concerning superposition applies.

The second thing to mention is that for each "event" (α-particle, β - particle or photon) there is a certain *detection probability* depending on the apparatus and the whole setup. Let the detection probability be p_{det}. It is assumed that the detection of a certain event does not affect the probability of detecting the next event. Let $(N(t))_{t>0}$ be the Poisson process we wish to study. Call the intensity λ. For $t > 0$, let $N_{\text{det}}(t)$ denote the number of events detected in $(0, t]$. It is the case that $(N_{\text{det}}(t))_{t>0}$ is again a Poisson process and the intensity λ_{det} is given by

$$\lambda_{\text{det}} = p_{\text{det}} \cdot \lambda.$$

This result is intuitively evident and is also easy to prove (see Exercise 24).

As a final complication, we mention that for most (all?) experimental setups a decay has the effect of causing a period of *"paralysis"* which prevents recording. We point out two models that include paralysis. Both involve a constant h, the *dead time.* Normally h is given in microseconds ($1\ \mu s = 10^{-6}$ second). The models are illustrated in Figure 6.

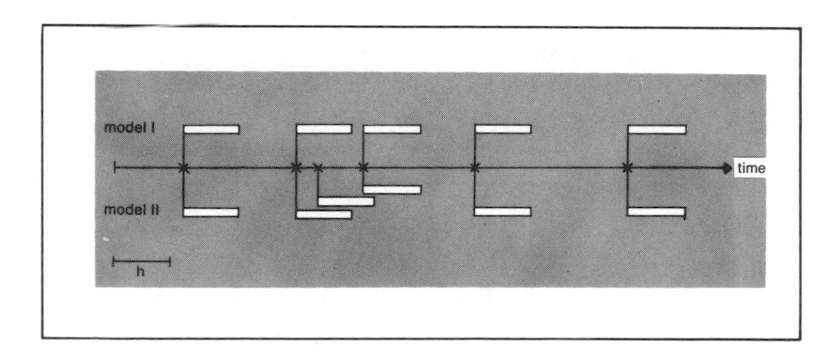

Fig. 6. Two models of paralysis due to the dead time h of the apparatus.

In both models, a decay registers only if it arrives at a moment when the apparatus is not paralysed, but for Model I only a recorded decay causes paralysis, whereas in Model II each decay (that reaches the counter) causes paralysis. For the situation in the figure there will therefore be 5 decays recorded under Model I and 4 under Model II.

The dead time arises from the whole measuring apparatus, both the counter and the associated electronics. However, it is often just the counter's dead time which comes into play. Whether Model I or Model II – or a blend of the two - is the correct one depends on the equipment. The difference between the two models is so small in any case (provided the dead time is much less than the mean waiting time) that we can choose to use the model which is mathematically easier to handle, if we do have to take the dead time into account.

In §16, analysing experiments carried out with school equipment, we shall come closer to the complications arising from the dead time, but the other points of criticism we have raised here will not play an appreciable role.

While the critique of mathematics in the previous section led to the desire for a better understanding of models (Thesis 18), we see that the points of criticism in this section lead us in a slightly different direction:

THESIS 19. *If one has been pleased with one model, one often has the desire to master other, more complicated models.*

This thesis is connected with the following:

THESIS 20. *Mathematics is an idealisation of reality.*

– the very fact that mathematics is idealised, and hence based on certain characteristics drawn from a limited number of situations, means there will always be a need for more refined models when further aspects and situations are discussed.

In connection with Thesis 19, note that the need for more models can also arise on purely internal, mathematical grounds.

Alongside with Thesis 20, one may also note that mathematics is by nature a basicly abstract science. The qualities of idealisation and abstraction formally puts mathematics at distance with the material world. It is, therefore, almost paradoxical that it seems to be these very qualities which guarantee mathematics its success, indeed make mathematics an indispensible tool in the study of the *real world* !

13. Other Applications of the Model

THESIS 21. *Good mathematical models have many concrete realisations.*

The formulation of properties (i), (ii) and (iii) in Theorem 1 is not restricted to radioactive processes. Thus there is a possibility that the Poisson process can be used as a model in other situations.

The considerations that led to setting up the Poisson process had their origin in some assumptions directly tied to the concrete phenomenon we were interested in. Assumptions are therefore not chosen from a desire to have as few preconditions as possible.

When one is interested in other applications of the model, it becomes interesting to know whether the model can be derived purely mathematically from fewer preconditions, or perhaps from completely different preconditions. This makes it easier to see, in concrete situations, whether the model can reasonably be applied.

Without giving a proof (which is difficult), we mention that the preconditions A_1 and A_4 suffice for the derivation of Theorem 1; assumptions A_2 and A_3 are therefore superfluous.

It is also worth mentioning that a Poisson process can be characterised by the fact that the waiting times are independent and exponentially distributed, with the same parameter. Here it must be pointed out that it is often very easy to see the reasonableness of the assumption that a random variable is exponentially distributed. Intuitively, one has to investigate whether the concrete phenomenon one sees is showing spontaneous behaviour. The important mathematical result which lies at the basis here is that the exponential distribution can be characterised by the equation (see Exercise 16)

$$P(v \geq t + s \mid v \geq s) = P(v \geq t).$$

Most of the examples of situations where the Poisson process gives a reasonable model are ones whose basic material is a "population." An "event" can be, e.g., "an individual from the population makes a telephone call," "an individual joins a queue," "an individual is involved in an accident", or the like.

14. The Poisson Approximation to the Binomial Distribution

The derivation we gave for Theorem 1 contained an unexpected bonus. The main point of the proof was in fact the following:

THEOREM 3. *A binomially distributed random variable with parameters (n, p), where n is large and p is small, has approximately the same distribution as a Poisson distributed random variable with parameter $\lambda = np$.*

The theorem is formulated very loosely. What we proved was that, for each $\lambda > 0$, and each $k = 0, 1, 2, \ldots,$

$$\lim_{n\to\infty} \binom{n}{k} \left(\frac{\lambda}{n}\right)^k (1 - \frac{\lambda}{n})^{n-k} = \frac{\lambda^k}{k!} e^{\lambda}.$$

This result does not in itself show how good the approximation is. One way to show this is by means of an example. We choose the case $\lambda = 5$. Table 2 shows the probabilities and cumulative probabilities (displaced) for a series of binomial distributions with $np = 5$ and for the Poisson distribution with $\lambda = 5$. The table was worked out with the help of program $P2$. The table shows that the Poisson approximation is good.

We mention without proof that the inequality

$$|P(X \in \Delta) - P(Y \in \Delta)| \leq \frac{\lambda^2}{n} \qquad (37)$$

holds for each subset Δ of $\{0, 1, 2, \dots\}$, for each $\lambda > 0$ and for each $n \in \{1, 2, \dots\}$, where X is binomially distributed with parameters $(n, \lambda/n)$ and Y is Poisson distributed with parameter λ.[9]

Table 2

	BINOMIAL DISTRIBUTIONS WITH np = 5				POISSON DISTRIBUTION
k	n = 20	n = 50	n = 100	n = 500	λ = 5
0	0.0032	0.0052	0.0059	0.0066	0.0067
	0.0032	0.0052	0.0059	0.0066	0.0067
1	0.0211	0.0286	0.0312	0.0332	0.0337
	0.0243	0.0338	0.0371	0.0398	0.0404
2	0.0669	0.0779	0.0812	0.0836	0.0842
	0.0913	0.1117	0.1183	0.1234	0.1247
3	0.1339	0.1386	0.1396	0.1402	0.1404
	0.2252	0.2503	0.2578	0.2636	0.2650
4	0.1897	0.1809	0.1781	0.1760	0.1755
	0.4148	0.4312	0.4360	0.4396	0.4405
5	0.2023	0.1849	0.1800	0.1764	0.1755
	0.6172	0.6161	0.6160	0.6160	0.6160
6	0.1686	0.1541	0.1500	0.1470	0.1462
	0.7858	0.7702	0.7660	0.7629	0.7622
7	0.1124	0.1076	0.1060	0.1048	0.1044
	0.8982	0.8779	0.8720	0.8677	0.8666
8	0.0609	0.0643	0.0649	0.0652	0.0653
	0.9591	0.9421	0.9369	0.9329	0.9319
9	0.0271	0.0333	0.0349	0.0360	0.0363
	0.9861	0.9755	0.9718	0.9689	0.9682
10	0.0099	0.0152	0.0167	0.0179	0.0181
	0.9961	0.9906	0.9885	0.9868	0.9863
11	0.0030	0.0061	0.0072	0.0080	0.0082
	0.9991	0.9968	0.9957	0.9948	0.9945
12	0.0008	0.0022	0.0028	0.0033	0.0034
	0.9998	0.9990	0.9985	0.9981	0.9980
13	0.0002	0.0007	0.0010	0.0013	0.0013
	1.0000	0.9997	0.9995	0.9994	0.9993
14	0.0000	0.0002	0.0003	0.0004	0.0005
	1.0000	0.9999	0.9999	0.9998	0.9998
15	0.0000	0.0001	0.0001	0.0001	0.0002
	1.0000	1.0000	1.0000	0.9999	0.9999
16	0.0000	0.0000	0.0000	0.0000	0.000
	1.0000	1.0000	1.0000	1.0000	1.0000

[9]Indication of proof: Let $(X_n)_{n \geq 1}$ and $(Y_n)_{n \geq 1}$ be sequences of independent identically distributed random variables, with X's Bernoulli distributed with parameter $p = \lambda/n$, and Y's Poisson distributed with parameter p. Put $S_n = \sum_1^n X_k$, $T_n = \sum_1^n Y_k$ and let δ_n be the supremum over Δ of $|P(S_n \in \Delta) - P(T_n \in \Delta)|$. Observe that $\delta_{n+1} \leq \delta_n + \delta_1$ (to compare S_{n+1} with T_{n+1}, first compare $S_{n+1} = S_n + X_{n+1}$ with $T_n + X_{n+1}$ and then compare $T_n + X_{n+1}$ with $T_n + Y_{n+1} = T_{n+1}$). It is easy to see that $\delta_1 = p(1 - e^{-p}) \leq p^2$. A somewhat more analytic proof starts by first noticing that the supremum defining δ_n is attained for Δ the set of k's with $P(S_n = k) > P(T_n = k)$; it follows that $\delta_n = \frac{1}{2} \sum_0^\infty |P(S_n = k) - P(T_n = k)|$.

The inequality (37) is qualitatively interesting, though not very sharp. Its main significance is that it indicates that it is not just the individual probabilities corresponding to particular k-values which agree well, but also the probabilities corresponding to several k-values.

Table 2 shows that the greatest numerical error we commit by replacing the binomial probability $P(X = k)$ by the Poisson probability $P(Y = k)$ is respectively 0.0268, 0.0095, 0.0045 and 0.0009 for the binomial distributions tabulated. For compound events the error can become larger, e.g.,

$$|P(4 \leq X \leq 7) - P(4 \leq Y \leq 7)| = 0.0260 \qquad \text{for } n = 50.$$

There are two ways, in particular, in which the Poisson approximation to the binomial distribution can be useful. One is in computation, where the Poisson probabilities are more accessible than the binomial probabilities, which are difficult to compute for large n. The other is in situations where one is still dealing with a binomial distribution, but without knowing the parameters n and p.

Let us look at an example. In L. von Bortkiewicz's book [3] of 1898 one finds the numbers in Table 3 of Prussian army corps that experienced $0, 1, 2, 3, 4$ or at least 5 deaths from horse kicks in the course of one year.

Table 3

number of deaths	number of army corps
0	109
1	65
2	22
3	3
4	1
≥5	0

This shows, e.g., that among the 200 army corps investigated there were 109 cases in which no death from a horse kick occurred during the year in question.

The army corps are stated to be of approximately equal size and composition, but the number n of soldiers in each corps is not stated (though by going to Bortkiwiecz's source, a Prussian statistical work, one would probably be able to discover the size of n).

Nevertheless, without knowing n we can still investigate a hypothesis about the nature of the phenomenon.

Let us try to explain the observed data on the assumption that the ·ent of a death does not affect the tendency of the survivors to allow themselves to be kicked to death. Thus we shall construct our model on the assumption that the tendency, in other words the probability, of a soldier dying from a horse kick in the course of a year, is the same for each soldier. We call this probability p.

Let X be a random variable to serve as a model for the number of deaths in an army corps by horse kicks in the course of a year. Our reflections on the model show that X must be binomially distributed with parameters n, p (there are n soldiers, who do not influence each other, and for each a probability p of "success").

The assumption underlying the model also shows that we have carried out 200 independent observations of the binomial distribution in question.

It is obvious that we are in a situation where we can replace the binomial distribution by a Poisson distribution to good approximation. Since the parameter of a Poisson distribution is just the mean, it is reasonable to estimate the parameter by the empirical average. This works out to be 0.61.

We can now compute the expected numbers of army corps with k deaths. This number is

$$200\frac{(0.61)^k}{k!}e^{-0.61}; \qquad k = 0, 1, \ldots .$$

Table 4

number of deaths	number of army corps, expected
0	108,7
1	66,3
2	20,2
3	4,1
4	0.6
≥5	0,1

Comparison of Tables 3 and 4 shows that the expected numbers agree well with the observed numbers. We conclude that our reflections on the model, which in all essentials treat the phenomenon as one of pure chance, are reasonable. More cautiously, we can say that on the above basis there is no reason to reject the model.

It is perhaps not out of place to mention that in reality the data came from 10, and not 200, army corps, but observed over 20 years (in the period 1875–1894). It would be sensible to expose this structure by a closer statistical analysis.

This book swarms with Poisson distributions, and some readers could come to the conclusion that all distributions have a natural connection with the Poisson distribution, so that the latter must be considered to be the most important one. Which distribution is the most important I shall leave unsaid, but it is of course only our choice of subject which places the Poisson distribution in an exceptional position. In the present section it should also be kept in mind that, in certain situations, the *normal distribution* (which we shall not go into further) can serve as an approximation to the binomial distribution, namely, when $np(1-p)$ is large. It follows that for λ large, the normal distribution can also serve as an approximation to the Poisson distribution. See, e.g., Feller [10], vol. I.

15. Spatially Uniform Distribution, Point Processes

A common situation involving a Poisson process is one we shall describe as a "spatially uniform distribution of particles". Imagine that particles are strewn evenly, but randomly, over a region of space. Here we normally think of either the usual 3-dimensional space or of the 2-dimensional space, the plane. However, one can also have the line (a 1-dimensional space) or higher dimensional spaces in mind.

Let us use a concrete example to illustrate the kind of phenomena that can occur. The example stems from a 1907 work of "Student" (pseudonym of William S. Gosset). It concerns the counting of yeast cells with a hæmocytometer, where a drop containing yeast cells is spread evenly in a thin layer between two object glasses. One object glass is divided into a large number of squares. It is then possible to count the number of yeast cells in each individual square.

Let N be the number of squares (in Student's case N was 400). The number of yeast cells in each square is regarded as a random variable. The number in the ith square is denoted by X_i; $i = 1, 2, \ldots, N$.

We aim to determine the stochastic nature of $X_1, X_2, \ldots, X_N$, which is another way of saying we wish to know the probability

$$P(X_1 = n_1,\ X_2 = n_2, \cdots,\ X_N = n_N) \tag{38}$$

for each finite sequence $n_1, n_2, \ldots, n_N$ of non-negative whole numbers. In particular, we want to know the distribution of the individual random variables $X_1, \cdots, X_N$.

The assumption we shall use to guide us can loosely be formulated by saying that particles are uniformly and randomly distributed. When we have found a model, then observations, held up against the model, can be used to test whether the assumption is reasonable and, if this is found to be the case, observations can be used to estimate the important parameters which go into the model.

We first remark that it seems reasonable to assume that $X_1, X_2, \ldots$ and X_N are independent (but also see the discussion below). If this is accepted, then we see that to be able to compute the probability written in (38), it suffices to know the distributions of the individual X_i's.

A necessary consequence of the assumption of uniform spatial distribution of particles and of the fact that all the N squares have the same area is that the X's have the same distribution. One can find this distribution in three different ways.

The way we shall follow first has as its starting point a fixed total number M of yeast cells in the N squares. From this it is easy to argue that X_1 must be binomially distributed with parameters $(M, n/N)$. Namely, we can think of X_1 being produced as the sum of M independent trial variables, each with a success probability of $1/N$; we merely have to let a success of the kth trial variable be the occurrence of the kth yeast cell in the first square ($k = 1, 2, \cdots, M$). If the conditions that permit the use of the Poisson approximation are satisfied, as will often be the case for the kind of phenomena we discuss here, then we get a model in which $X_1, X_2, \ldots, X_N$ all have the Poisson distribution with parameter M/N.

Combining this result with the independence of the X_i's gives a complete model. But there is an essential objection that occurs at this point, namely, that the X_i's cannot possibly be independent, since $X_1 + X_2 + \cdots + X_N = M$. E.g, we must have

$$P(X_1 = 0, \cdots, X_N = 0) \neq P(X_1 = 0) \cdots P(X_N = 0).$$

Here it can be remarked that, although the X_i's cannot be exactly independent, they must surely be in some sense approximately independent. This argument is perhaps not particularly convincing, and we shall now give an alternative line of reasoning.

This new approach considers the total number of yeast cells, M, not as a fixed number, but as a random variable. This is a realistic assumption, since if we take another drop containing yeast cells and carry ou observations on it, then we naturally cannot expect the total number of yeast cells in the N squares to be the same.

We also assume that the conditional distribution of the X's, given that M takes a particular value M_0, is a binomial distribution with parameters $(M_0, 1/N)$. Finally, we have to make an assumption about M's distribution. With the background we now have, the reader will perhaps accept the assumption that M is Poisson distributed. It can then be shown that the X's remain Poisson distributed (see Exercise 27).

The above line of argument is not completely satisfying either, since the assumption about M's distribution was rather unmotivated. The last approach we shall suggest will, we hope, be more convincing to the reader.

We shall now proceed from the assumption that to *each* region of the object glass there corresponds a random variable which gives the number of yeast cells in the region. Thus we are now looking for a stochastic model which is much more refined, since we want to characterise a whole family $X(\triangle)$; $\triangle \subseteq G$ of random variables, where $\triangle$ varies over all subsets $\triangle$ of the object glass G (see Figure 7).

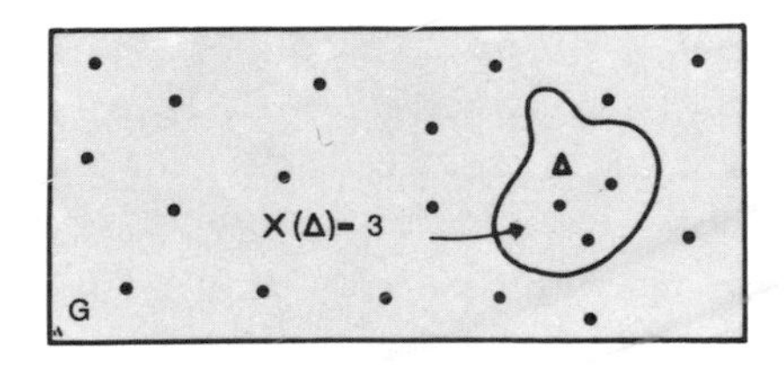

Fig. 7. Object glass with subarea containing 3 yeast cells.

The conception we have of the yeast cell distribution leads us to make the following assumptions:

B_1. *If* $\triangle_1$ *and* $\triangle_2$ *have the same area, then* $X_1(\triangle)$ *and* $X_2(\triangle)$ *are identically distributed.*

B_2. *If* $\triangle_1$ *and* $\triangle_2$ *are disjoint, then* $X(\triangle_1)$ *and* $X(\triangle_2)$ *are independent and* $X(\triangle_1 \cup \triangle_2) = X(\triangle_1) + X(\triangle_2)$.

If we now want to find the distribution of $X(\triangle)$, where $\triangle$ is a square, then we can subdivide $\triangle$ into small squares $\triangle_1, \ldots, \triangle_\nu$ and view $X(\triangle)$ as the sum of $X(\triangle_1), \ldots, X(\triangle_\nu)$. Now we recognise the train of thought which led to Theorem 1! We shall therefore not go into detail, but merely conclude that with an extra regularity assumption one can show that the X's satisfy

B_3. *There is a constant* λ *such that* $X(\Delta)$ *is Poisson distributed with parameter* $\lambda \cdot \mid \Delta \mid$, *where* $\mid \Delta \mid$ *denotes the area of* Δ.

The properties B_1, B_2 and B_3 determine the stochastic nature of the random variables $X(\Delta)$. We say that the $X(\Delta)$'s constitute a *point process of intensity* λ *over the region* G. The intensity is therefore the average number of particles per unit area.

We find that the three lines of argument lead to the same model for yeast cell observation: the random variables that describe the number of yeast cells in the individual squares are independent and Poisson distributed. The parameter that goes into the Poisson distribution can be estimated from the observed average number of yeast cells per square.

Of course, one must see whether the modelling is confirmed by actual observations.

Table 5 shows Student's material. There are 4 sets of observations, each of which corresponds in the model to observation of 400 independent Poisson distributed random variables. The parameters are estimated respectively as 0.6825, 1.3225, 1.80 and 4.68. The expected numbers of squares containing k yeast cells can now be computed. These numbers are also shown in Table 5. It should be mentioned that the expected number given last for each set of observations corresponds to the event "k or more."

Table 5

	I		II		III		IV	
k	obs.	exp.	obs.	exp.	obs.	exp.	obs.	exp.
0	213	202.1	103	106.6	75	66.1	0	3.7
1	128	138.0	143	141.0	103	119.0	20	17.4
2	37	47.1	98	93.2	121	107.1	43	40.6
3	18	10.7	42	41.1	54	64.3	53	63.4
4	3	1.8	8	13.6	30	28.9	86	74.2
5	1	(≥) 0.3	4	3.6	13	10.4	70	69.4
6			2	(≥) 1.0	2	3.1	54	54.2
7					1	0.8	37	36.2
8					0	0.2	18	21.2
9					1	(≥) 0.0	10	11.0
10							5	5.2
11							2	2.2
12							2	(≥) 1.3

If readers wish to see how the theory works in their own experiment, this can easily be done as follows. One simply throws small particles (sesame seeds, grains of sand or the like) over a surface, e.g., a floor or a table top. A suitable part of the surface is divided into subregions of equal size, and one makes a count of these as above. The count can be compared with the theoretically expected numbers, exactly as in the treatment of Student's results.

16. A Detailed Analysis of Actual Observations

The main aim of this section is to carry out an analysis of data from an experiment with a radioactive source using only equipment of the type which today is found in every university or college. The experiment we describe can even be carried out in high school.

The analysis, which to many will certainly seem surprisingly long, falls into the following parts. First, a mathematical model is set up and tested. This shows that the model must be rejected. A modified model, which also takes account of the observing equipment, is then set up, tested and, it turns out, accepted.

We take the opportunity to mention some general statistical considerations. One can attach more or less importance to these, according to one's interest in modelling. There is a certain freedom in the treatment, which allows personal taste to be expressed. However, the technical means at our command will always set limits on what it is possible to do. Thus for certain investigations it will be necessary to have access to a programmable calculator or, even better, a personal computer.

Experimental Protocol.

Experimenter: Malte Olsen, Physical Laboratory I, University of Copenhagen.

Date: A day in June 1981.

Counter:

Type and age: Geiger-Müller tube, c. 10 years old.

Maker and specifications: 20th Century Electronics B6 HL.

Voltage: 640 V.

Dead time for new counter, given by the manufacturer: 100 μs (= 0.0001 sec.).

Background: 1942 counts in 13 minutes, i.e. c. $2\frac{1}{2}$ counts per sec.

Two-source measurements: Two β-sources from RISO were used. Two experiments were carried out. In each experiment, on each preparation, three counts during 10 seconds were carried out.

	Experiment I			Experiment II		
Source 1	3810	3866	3834	3013	3059	3027
Source 1 + 2	4469	4459	4439	3852	3857	3870
Source 2	3722	3731	3706	2931	2952	2986

The observations we shall study in this section are reproduced in Table 6. The source used was the β-active source specially prepared for educational purposes by the RISO National Laboratory, in conjunction with the Union of Physics and Chemistry Teachers of High Schools and Colleges[10]. The production of β-particles occurs with the following pro-

[10]The radioactive sources provided by RISO are easy to work with and – a main consideration – safe in the hands of non-experts. The sources are available for experiments performed by pupils even in elementary school and have been released for distribution by health authorities in a number of countries including the Scandinavian ones. In the US it is also possible to perform simple experiments as the one decribed, but outside the research institutions this is more difficult as the appropriate radioactive sources have to be obtained on a case by case basis from the Oakridge National Laboratory. *Information provided by Kaj Heydorn, RISO.*

Main experiment: 249 counts for a β-source from RISO. Time interval 1 sec.

Table 6

81	75	83	78	79	60	75	77	60	71	90
85	72	73	80	66	81	73	76	79	81	72
79	82	75	75	79	75	79	69	83	78	71
75	74	77	74	73	80	83	74	68	76	72
73	73	80	84	76	63	74	84	75	76	80
70	72	76	74	75	73	72	61	78	76	78
77	78	83	65	74	72	74	71	83	78	82
86	72	67	68	81	65	78	90	67	68	80
68	70	67	69	79	71	69	82	74	88	75
69	67	82	70	78	75	82	81	76	74	67
63	70	78	80	90	68	69	75	71	78	78
66	78	74	68	73	74	75	66	74	69	70
81	84	78	63	89	70	79	94	89	75	73
71	62	72	83	78	100	77	82	76	69	80
72	77	82	77	81	74	68	77	79	77	75
64	70	73	66	72	69	86	73	63	75	85
69	83	71	72	79	75	83	82	69	73	83
76	68	94	76	82	74	77	81	74	78	77
68	72	79	74	85	72	71	86	88	84	81
85	80	76	72	71	72	75	72	82	77	
74	84	65	65	69	67	71	64	76	77	
72	75	77	74	76	76	94	77	81	85	
78	73	73	75	60	70	88	82	74	71	

---

cesses, a *mother process and a daughter process*:

$$
\begin{aligned}
{}^{90}_{38}Sr \rightarrow {}^{90}_{39}Y + {}^{\;\,0}_{-1}e \qquad & \text{(mother process; half-life 28 years)} \\
\hookrightarrow {}^{90}_{40}Zr + {}^{\;\,0}_{-1}e \qquad & \text{(daughter process; half-life 64 hours).}
\end{aligned}
$$

Had there been only one process (with a half-life that was large in comparison with the duration of observation), then it would have been completely natural to use a Poisson model, i.e., we could safely have claimed that the process in which an "event" corresponds to emissions of a β-particle from the source is a Poisson process. The same is also true, however, when we take account of the fact that there is both a mother and daughter process!

And the reason for this is not that we only get a few β-particles that stem from the daughter process. In fact, half the β-particles emitted stem from the latter process (see Exercise 42!)[11] The essential point, rather, is that the process can be regarded as the superposition of two independent Poisson processes (and this yields a Poisson process, as mentioned in §12 and Exercise 25). This state of affairs holds partly because the time the experiment lasts (c. 5 minutes) is much less than the half-life for both the mother and daughter processes, and partly because the half-life for the daughter process is large enough to avoid dependence (if the half-life for the daughter process had been much smaller there would have been strong dependence, inasmuch as each β-particle stemming from the other process would have been accompanied by an immediate successor β-particle from the daughter process).

It should perhaps be mentioned that the considerations above apply in a way to the whole source, whereas the arrangement and construction of equipment which surrounds the source allows only a minority of the β - particles produced to enter the counting chamber. Thus each β-particle has a certain probability of being counted. This state of affairs does not ruin a Poisson process, however, it obviously just entails a reduction of intensity (see §12 and Exercise 24).

To illustrate some of the effects we have mentioned, we point out that during the experiment (duration c. 5 min.) c. 0.2 per 10^6 of the Sr atoms which were in existence at the beginning of the experiment are transformed, and c. 1 per 10^3 of the Y atoms are transformed.

If one feels insecure about the presence of two processes and the possibility of this ruining the "Poisson character" of the experiment, then one might try to filter out the low energy β-particles from the mother process. However, this is not without problems, since strong filtration produces an accompanying braking radiation which causes interference, and hence one must take account of the fact that β-particles do not have well-defined energy, but instead exhibit a continuous energy spectrum (as is well known, this state of affairs is connected with the fact that a neutrino is emitted at the same time as the β-particle; the latter will certainly not be recorded by standard equipment and hence does not represent a further complication).

One can also seek to avoid the problem completely by working instead with the α-active RISO source, where the process is

$$^{241}_{95}Am \rightarrow {}^{237}_{93}Np + {}^{4}_{2}He \quad \text{(half-life 470 years).}$$

[11]Since the β-particles from the daughter process have much higher energy (maximum energy) than those from the mother process, it is even the case that most (c. 80%) of the β-particles which escape the equipment surrounding the source are from the daughter process.

But alas! For this process we have to admit that there is a weak γ - radiation, so here again another sceptical soul will perhaps feel the need for a more precise investigation.

It is not at all easy to arrange an experiment so that the Poisson model, which we have taken such pains to set up, can be said without hesitation to be the completely obvious one!

THESIS 22. *Those who seek a phenomenon which exactly follows a mathematical model, seek in vain.*

This is also related to Thesis 20.

Let us summarise the results of the above discussion: even though one can imagine an experiment carried out under more ideal conditions, the disturbing factors we have pointed out play only a small role. We therefore expect that the process $(N(t))_{t\geq 0}$, where $N(t)$ is the number of β-particles which arrive in the counter in the time interval $]0, t]$, would be very nearly a Poisson process (at any rate for $t < 5$ minutes, the approximate duration of the experiment). Consequently, we expect that the 249 observations under consideration can, with high accuracy, be regarded as independent observations from a Poisson distribution.

We shall dwell a little longer on these reflections on our model, in order to become quite clear about which notions we wish to validate or invalidate with our experiment. Our expectations of the process $(N(t))_{t\geq 0}$ are due, as was pointed out above, to the fact that we reckoned the process $(N_{\text{total}}(t))_{t\geq 0}$ to be a Poisson process, where $N_{\text{total}}(t)$ denotes the total number of β-particles emitted by the source in $]0, t]$. If we think back to our basic discussion of the Poisson process (and the circumstances brought to light in Exercises 16, 24 and 25 should also be mentioned) then we see that the essential basis of these expectations, in brief, was the supposed *spontaneity* of radioactive nuclear transformations. In other words, such transformations, we believe, are nondeterministic – stochastic in principle.

The aim of our experiment can therefore be said to be investigating whether the idea of spontaneous nuclear transformation is correct. A complete confirmation is obviously unattainable. We cannot directly investigate an entire Poisson process. Instead we have chosen to carry out a large number of observations over equal, non-overlapping time intervals. If we find that the observations behave like independent observations from a Poisson distribution, then we shall take this as some indication that our modelling holds good.

Let us now apply ourselves to the above data. In the terminology introduced in Exercise 17, we can say that the above data is a *sample* of size 249, and we conjecture that the sample is from a Poisson distribution. This conjecture will be investigated. This is a typical statistical problem.

THESIS 23. *The aim of a statistical investigation is to proceed, from given observations of a stochastic phenomenon, to draw conclusions about the underlying model as a means to "explain" the phenomenon.*

It is important to keep in view that the starting point of a statistical analysis is an actual observation, and hence a finite set of numbers. Therefore, one cannot normally expect to arrive at definitive conclusions.

THESIS 24. *Statistical investigations lead to decisions which are taken under uncertainty.*

A very important type of statistical problem arises when a sample is given and one wishes to decide which distribution the sample is taken from. Let us look more closely at the present sample. As a beginning, we shall just endeavour to represent the numerical material in a convenient, surveyable way.

An obvious possibility is to represent the numbers in a "tally diagram", which is essentially the same as a histogram (see Table 7, rotate 90° anticlockwise).

From this diagram one can directly read off the number N_k of observations in which k β-particles were observed. The numbers N_k , which in contrast to the case of Rutherford and Geiger's counts (see §10), are relatively small, may also be illustrated by the so-called *empirical distribution function*, the meaning of which is evident from Figure 8 (the curve drawn in black). When one reads the number 160 above the 77 mark, this means that in 160 out of the 249 observations at most 77 β-particles were observed. The actual values of the empirical distribution function have been normalised; thus the value of the empirical distribution function for argument 77 is not 160, but $160/249 = 0.64$, as can be read off the scale given in parentheses. The normalisation means that the empirical distribution function actually is a distribution function. The corresponding distribution is called the *empirical distribution* (corresponding to the sample). One often refers to the values of the empirical distribution function as the *cumulative frequencies.*

As in §10, we let U and V denote the number of time intervals and the total number of observed events respectively. In addition, we set

$$T = \text{ length of a time interval}$$

and

$$\lambda_{\text{reg}} = \frac{V}{UT}, \tag{39}$$

which is the *registered intensity* measured in counts per unit time.

For the actual data we have

$$T = 1 \text{ sec.}, \quad U = 249, \quad V = 18801, \quad \lambda_{\text{reg}} = 75.5,$$

(the units for λ_{reg} are sec.$^{-1}$).

We shall compare the empirical distribution with the Poisson distribution which has parameter λ_{reg}. We know from Exercise 17 that λ_{reg} in a certain sense is the most reasonable value we can ascribe to the parameter; λ_{reg} is its *maximum likelihood estimator*. In other words, if we compute the probability of getting precisely the observed sample for different values of the parameter λ then, basing the calculation on the Poisson distribution with parameter λ, we find that this probability is greatest when $\lambda = \lambda_{\text{reg}}$.

We graph the distribution function of the Poisson distribution with parameter λ_{reg} as the grey curve in Figure 8. In this way it becomes evident that with the scale used, where the probability is multiplied by 249, and provided the sample is from a Poisson distribution with parameter λ_{reg}, we should expect that in 149 out of 249 observations there will be 77 or fewer β-particles. This follows from the stabilisation of relative frequencies (theoretically, the law of large numbers).

Since the Poisson distribution we have arrived at obviously departs noticeably from the empirical distribution, we must conclude that the Poisson model is not suitable. There is no possibility of maintaining the assumption of a Poisson distribution. It is certainly suitable as an approximation, but the departures point to a *systematic* error, since the Poisson distribution function clearly lies above the empirical distribution function for small counts, and below it for large counts.

There is therefore no reason to make a complicated statistical investigation to see that the model is not suitable – we merely have to look at the two distribution functions in Figure 8.

THESIS 25. *Statistics is an aid, and not a substitute for common sense.*

Table 7 "Bier diagram"

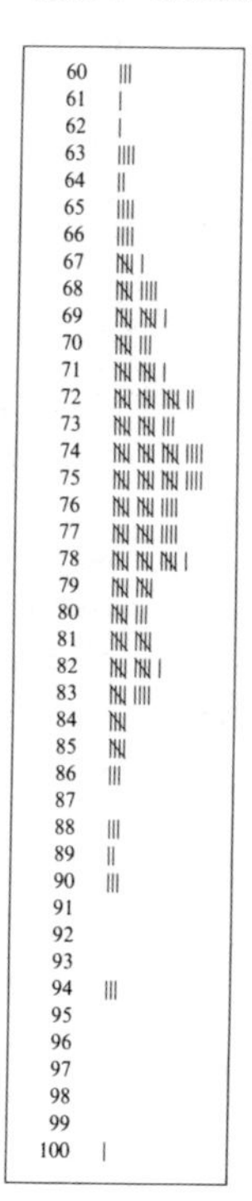

60				
61		
62		
63					
64			
65					
66					
67	𝍸	
68	𝍸 ||||
69	𝍸 𝍸	
70	𝍸 |||
71	𝍸 𝍸	
72	𝍸 𝍸 𝍸 ||
73	𝍸 𝍸 |||
74	𝍸 𝍸 𝍸 ||||
75	𝍸 𝍸 𝍸 ||||
76	𝍸 𝍸 ||||
77	𝍸 𝍸 ||||
78	𝍸 𝍸 𝍸	
79	𝍸 𝍸
80	𝍸 |||
81	𝍸 𝍸
82	𝍸 𝍸	
83	𝍸 ||||
84	𝍸
85	𝍸
86				
87	
88				
89			
90				
91	
92	
93	
94				
95	
96	
97	
98	
99	
100		

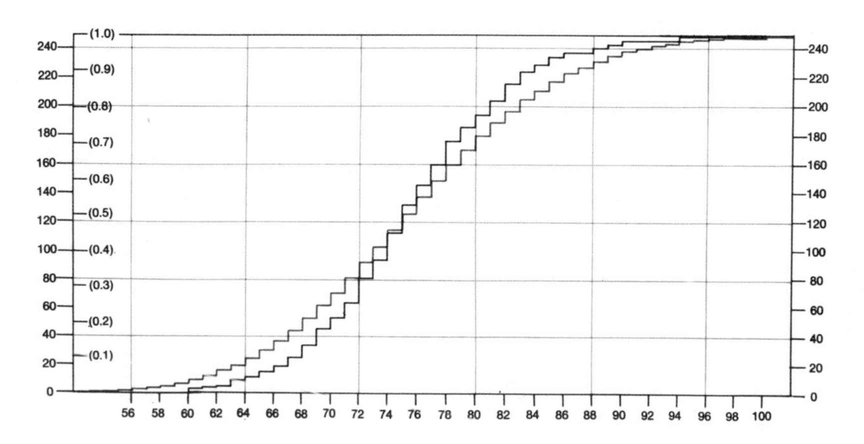

Fig. 8. Approximation by a Possion distribution (grey) to the empirical distribution function (full-drawn).

After the basic discussion in §12 we are prepared to meet with a situation like this, where a model must be thrown out. We also know how it might be possible to fix it up. We elect to refine our considerations by taking account of the dead time of the counter (and its electronics).

The dead time we now introduce into the discussion arises from the Geiger-Müller tube. Let us take the opportunity to give a rough account of the way a Geiger-Müller tube works, in order to understand how the dead time comes about.

For the Geiger-Müller tube to work, it is necessary to impose a potential difference between the walls of the tube (negatively charged) and a thread through the axis of the tube (positively charged). When a β-particle, which is the subject of our experiment, comes into the tube, ionisation occurs. There is an avalanche effect which causes a host of electrons to arrive at the thread for c. 10^{-4} sec., and gives rise to a sudden voltage drop which, with the help of suitable electronic equipment is amplified and leads to the registration of an individual "event". It takes a somewhat longer time, c. 10^{-4} sec., for the positive ions produced to arrive at the walls of the tube. In this time interval the positive charge prevents the registration of possible new β-particles and the counter is "dead".

Let us set up a model that takes account of the dead time. We retain the primary assumption that $(N(t))_{t\geq 0}$ is a Poisson process with (unknown) intensity λ. But $N(t)$ is the total number of β-particles that arrive in the counter, and we let $N_{\text{reg}}(t)$ denote the corresponding number of registered β-particles. We let $P_{\lambda,h}$ denote probabilities computed on the assumption that the dead time is h and the intensity of $(N(t))_{t\geq 0}$ is λ.

We have the following formula, which is proved in Exercise 37.

$$P_{\lambda,h}(N_{\text{reg}}(t) \leq n) = e^{-\lambda(t-nh)} \sum_{k=0}^{n} \frac{(\lambda(t-nh))^k}{k!}. \tag{40}$$

A precondition for its validity is that $nh < t$.

If λ, h and t are fixed, then (40) defines a certain distribution function and hence a certain distribution. We refer to this distribution as the *modified Poisson distribution.*

The new model we have arrived at in this way proceeds, like the old one, from the assumption that $(N(t))_{t\geq 0}$ is a Poisson process. But, in contrast to the old model, it takes account of the dead time. This implies that we now expect the above sample to be from a modified

Poisson distribution with parameters λ (= intensity of $(N(t))_{t\geq 0}$) and h (= dead time) and T. Since T is known (= 1 sec.), the corresponding cumulative probability can be computed (by machine!) from (40), as soon as λ and h are known. See Program $P4$. Also note that $h = 0$ gives the usual cumulative Poisson probabilities.

There now remains the essential problem of determining λ and h from observations, or rather of determining good *estimators* of these quantities.

The latter problem can be tackled in many ways. Let us first see what we get by applying the maximum likelihood method. The idea is basically simple, along precisely the same lines as Exercise 17, where a single parameter remains to be estimated. The computation is difficult from the technical viewpoint, but fortunately we have the beloved computer to save us the pain!

The likelihood function $L(\lambda, h)$ is now a function of the two parameters λ and h, and it is given as the product, over all possible values of k, of the quantities

$$\{P_{\lambda,h}(N_{\text{reg}}(T)) = k\}^{N_k}.$$

Think it over! For the actual set of observations k runs through values from 60 to 100. The latter function is to be optimised. This can be achieved in principle by a suitable program for computing $L(\lambda, h)$, or better $-\log L(\lambda, h)$, and simply feeling one's way. But here we are confronted with a problem that is so extensive that it cannot be carried out on many of the calculators we can think of using.

We can therefore ask whether we are not in a favourable situation similar to that of Exercise 17, where the optimisation problem could be solved explicitly. In other words, we ask whether we can find a formula which directly expresses the maximum likelihood estimators of λ and h in terms of observed quantities (the numbers N_k together with T). The answer is "No – but almost!" One can show that the desired values of λ and h lie very close to the values of λ and h one gets by solving the two equations

$$\lambda = (\lambda_{\text{reg}}^{-1} - h)^{-1}, \tag{41}$$

$$\sum_k \frac{k(k-1)N_k}{T - kh} = \lambda \sum_k kN_k \quad (= \lambda V). \tag{42}$$

In (41) we recognise formula (51) from Exercise 35. The latter formula, whose content is easy to understand (see Exercise 35), can be used to reduce the foregoing optimisation problem in two variables to an optimisation problem in one variable.

I shall not try to account for formula (42). If the reader is willing to accept it without further explanation, then I recommend that it be used together with (41) in the following way: one substitutes the expression for λ in (41) on the right-hand side of (42). One can then calculate the difference between left- and right-hand sides of (42) for various values of h and try to find a value for which the difference is 0 or so close to 0 that one is satisfied. The value of h found in this way can then be substituted in (41) to find λ. Program $P3$ may be used for this purpose. To test whether the parameter set (λ, h) found really is the maximum likelihood estimator, or close to it, we compute $-\log L(\lambda, h)$ together with $-\log L(\lambda_1, h_1)$ and $-\log L(\lambda_2, h_2)$, where h_1 is a little smaller, and h_2 a little larger than h, and λ_1 and λ_2 are determined by (41). That should readily show that $-\log L(\lambda, h)$ is the least of the three values computed. To gain full confidence, one ought to compute another six values, corresponding to λ-values a little above and below the values just mentioned. The computations may be carried out via Program $P5$. Depending on the computing equipment available, the suggested test may take some time.Those who have faith in (41) and (42) can naturally skip the tests. Proceeding as indicated, one finds

$$h = 0.00310 \text{ sec.} = 3100\mu s, \quad \lambda = 99 \text{ sec.}^{-1}. \tag{43}$$

But this is not the end of our investigation. We must see whether the modified Poisson distribution with the parameters in (43) describes the observed material better than the Poisson distribution we saw in Figure 8. And luckily, it does! After the rather lengthy statistical analysis we are rewarded by a fine agreement between the empirical distribution function and the distribution function of the modified Poisson distribution, see Figure 9 (computed with the help of Program $P4$).

We can conclude from this that the observations do not give cause to abandon the central assumption about the phenomena in question that it is of a principally stochastic nature, a truly spontaneous phenomenon. Indeed, it is tempting to take the results of the statistical analysis in support of the latter view.

There is also a certain aesthetic satisfaction to be had from the modification of our model. The model is clearly divided into two parts, one that handles the phenomenon we are studying, and one that takes account of the disturbances introduced by the method of observation chosen. And of the two parameters we work with, one exclusively concerns the phenomenon itself, and the other exclusively concerns the observational equipment.

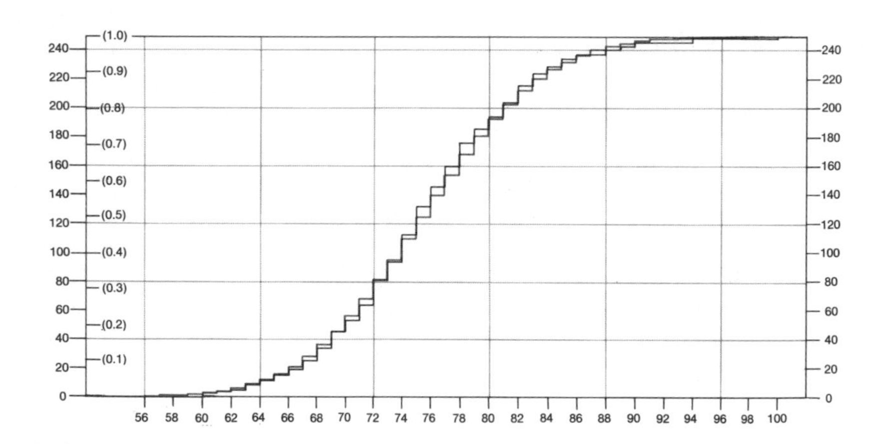

Fig. 9. Approximation by a modified Poisson distribution (grey) to the empirical distribution function (full-drawn).

Ever since we realised from a glance at Figure 8 that the Poisson model was not usable, it would have been easy to introduce an "ad hoc parameter" (e.g. expressing steepness of the distribution function) and thereby modify the model so that better agreement was obtained. However, it is a central point of our analysis that the extra parameter we introduced has a direct meaning in relation to the experiment we carried out. One should always try to ensure that the parameters introduced into a model have such a clear interpretation. In addition, the parameters should describe different, completely separate aspects of the phenomenon under consideration. This also limits the number of parameters.

In many models, and I have some fishery and economic models particularly in mind, one operates with so many "ad hoc parameters" that the models are "doomed to success" – one can easily get them to fit the observations made. Such models nevertheless have their uses, e.g. in working out prognoses, but one must realise that first and foremost they are descriptive and they do not in any great measure yield a proper understanding of the underlying mechanism.

As for the two parameters λ and h, it should also be remarked that the precise value of λ is not really of interest. The intensity λ in fact also concerns the counter and not what is more interesting, the source itself. The true intensity of the source cannot be obtained from λ without further calculation. The latter requires knowledge of the geometry and properties of the materials; the reason for this is that the passage of the radiation through air, glass, etc. involves a weakening, an *attenuation* (see Exercise 46).

As far as the source is concerned, we maintain that it is the purely qualitative aspects that we are interested in - the spontaneous nature of radioactivity. On the other hand, knowledge of the other parameter h gives important quantitative information about the observational equipment.

At the time we gave up the original pure Poisson model we chose to introduce the dead time. In fact, that is the only possibility which is able to explain – note *explain* and not merely describe – the large departure from the Poisson model, and it happens very frequently, for the type of data set we are dealing with here, that one is obliged to take account of the dead time. Thus it may be mentioned that an apparently abnormal (non-spontaneous) behaviour of observations of cosmic radiation was explained completely satisfactorily just by taking account of the dead time.[12]

It is easy to see, purely qualitatively, whether the departure of a given data set from a pure Poisson model stands a chance of being explained by the introduction of the dead time. If the empirical distribution function and the Poisson distribution function look qualitatively like Figure 8, then there is a possibility of doing so. This can be seen, very roughly, by observing that when the dead time is very large in relation to the mean waiting time the distribution function for the number of registered decays approaches more and more closely to a very steep function which grows suddenly from 0 to 1. Think it over!

Even though our investigations gave a nice result, summed up by Figure 9, there are nevertheless grounds for criticism. Namely, it is rather conspicuous that the h-value we arrived at is much bigger than expected. The manufacturer gave a value c. 30 times smaller! Here we should note the following. The given value of 100 μs corresponds to ideal working conditions, a certain count per second and an optimal voltage. Naturally one aims to carry out the experiment under these conditions, but small departures from them can mean that the dead time becomes quite large. A factor of 30, however, can certainly not be

[12]See Levert and Scheen in *Physica* X 1943 and Feller's article in the *Courant Anniversary Volume*, 1948.

explained in this way.

This much larger departure is due to the fact that the given dead time is valid for new apparatus, while the actual dead time increases appreciably with time. We can suggest reasons for this after looking a little more closely at the working of the Geiger-Müller tube. We have already remarked that the dead time is due to the positive ions formed, which take a relatively long time, c. 10^{-4} sec., to arrive at the wall of the tube. There is, however, a risk that the ions, on arriving at the wall, will produce new ions, which will prolong the dead time. To hinder this process there is an *extinguisher*, usually a halogen, which absorbs energy from the positive ions and thereby hinders further ionisation. With age, the halogen will gradually be "consumed" by diffusion of its molecules into the metal electrodes (diffusion can possibly also occur through the thin window of the tube). In this way the ionisation and the dead time can be prolonged.

Experience shows that one should not be surprised to find dead times which differ by a factor of up to 10 from the one given by the manufacturer. A factor of 30, such as we found, is too much and actually, the experiment led to the counter being discarded (a closer examination showed that the dead time found corresponds quite closely to the dead time one expects to find in a tube completely empty of extinguisher gas!).

We shall also mention an easy and popular method which gives an independent determination of the dead time. The method goes under the name of the *two-source method* and it is based on the fact that the dead time comes into play more strongly as the intensity increases. The method is treated in detail in Exercise 35, to which we refer. Applying the method to the observations in the experimental protocol, we get $h = 1850\ \mu s$ (for Experiment I we get $h = 1851 \mu s$ and for Experiment II $h = 1849\ \mu s$; of course it is a coincidence that the two values of h lie so close together).

In summary, we can say the following about the experiment treated here: one must be prepared to take account of the dead time. The dead time given by the manufacturer is seldom relevant. An independent determination of the dead time by the two-source method should always be carried out. If one has a computer at disposal, then it is a good idea to determine the dead time directly from observations, by the maximum likelihood method. The two determinations of dead time are often quite inaccurate, and can lead to rather different values. To obtain a good determination of the dead time, one can seek a reduced intensity in the two-source method, and work with an intensity of the same order of magnitude as that for the main experiment (this was not done in our experiment); here the background comes into play more strongly and the lower intensity means on the whole that one has to use longer observation times in the two-source method.

More observations in the main experiment also lead to the dead time determined by the maximum likelihood method becoming more accurate.

In carrying out the observations, one can risk finding the dead time to be 0. Naturally this can be because one is working with a counter with an exceptionally short dead time (a brand new tube or a very fast type, such as, e.g., a solid state counter or a photomultiplier), and in such a case a dead time of 0 is fully acceptable.

The above remarks are purely qualitative and of course one can ask whether it is possible to come up with quantitative statements which can assist in the planning of experiments. Here I refer to Exercises 34 and 36 which show, on the one hand, how the main experiment should be organized so that λ is accurately determined and, on the other hand, how the two-source measurements should be organised so that h is accurately determined. It is difficult to organise the main experiment so that one can be sure in advance that h, computed by the maximum likelihood method, will be accurately determined

It is hoped that the discussion in this chapter will help readers to organise their own experiments. The aim has also been to demonstrate the power of a detailed statistical analysis. It should be clear that the analysis was not exhaustive. E.g., we could submit our data, where the sequence of individual observations was taken down, to statistical tests for independence. However, we shall not go further into this. On the other hand, in the next chapter we shall see a general statistical method of great scope.

It is tempting, and also relevant, to conclude with a remark on the classical experiment of Rutherford and Geiger. Here we have no possibility of determining the dead time by the two-source method, and the maximum likelihood method is therefore of special interest. It is clear that there must have been a dead time for this experiment, and this was also pointed out by Rutherford and Geiger. The light impression received by the eye with a microscope observing a scintillation on a scintillation screen – and that was the method of observation used here – needs a little time to die down and to allow a new scintillation to be registered. Certainly it is more reasonable to use Model II of §12 as the dead time model, but one can use Model I as an approximation, since it is mathematically far simpler to handle. A calculation gives the value $h = 0.045$ sec., which agrees well with what we would have expected in advance. Think it over!

17. The chi–square Test for Goodness of Fit

In the foregoing we have often combined empirical and theoretical material (Table 1 and the associated Figure 4, Tables 3, 4 and 5, Figure 8 and Figure 9). And each time the connection was visible to the eye, so to speak. This is unsatisfactory in the long run, and the need arises for more objective criteria.

We shall describe a noteworthy method that meets this need. The method is not specific to any particular type of distribution, but is completely general. We have to be content with a description of the method that is merely enough to enable the reader to use it. A deeper understanding of why the method works cannot be given here.

The method is based on a family of distributions, the so-called χ^2-*distributions* ("chi-squared distributions"). A distribution in this family is defined for each natural number r. The distribution corresponding r is called the χ^2-*distribution* with r *degrees of freedom* and it is defined to be the distribution with support $[0, \infty)$ which is characterised, up to a constant factor, by the density function $x \mapsto x^{r/2-1} \cdot e^{-x/2}$; $x > 0$.[13]

Figure 10, which shows the density functions for a few χ^2-distributions, gives an idea of the way the distribution varies with the number of degrees of freedom. However, the information contained in Figure 11 is of greater practical value.

[13]The distribution can also be characterised as the distribution of the random variable $X_1^2 + X_2^2 + \cdots + X_r^2$ where X_1, $X_2, \ldots, X_r$ are independent and normally distributed with mean 0 and variance 1.

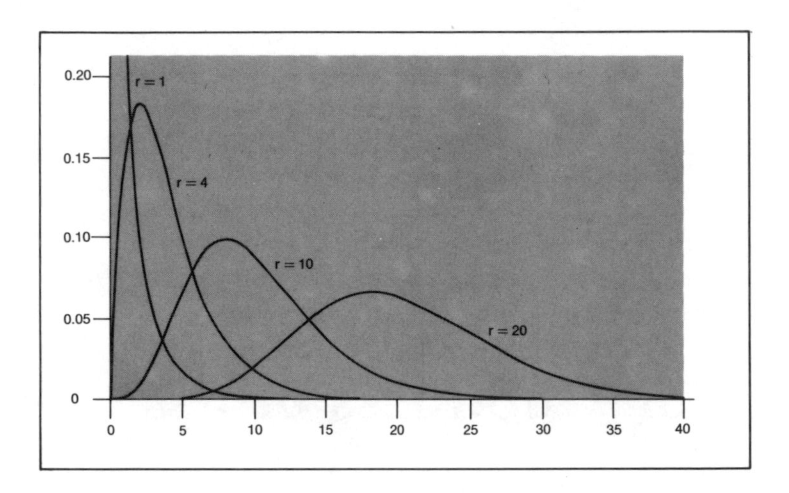

Fig. 10. The density functions for the χ^2-distributions with, respectively 1, 4, 10 and 20 degrees of freedom.

Here we can find the probability in % that a given χ^2-distribution assumes a value greater than a given number. E.g., we see, since the point with coordinates (43,27) lies between the 1% and 5% curves, that $P(X \geq 43) \approx 3\%$ when X has the χ^2-distribution with 27 degrees of freedom. We use the notation

$$P(\chi^2_{27} \geq 43) \approx 3\% \tag{44}$$

to express this. Similarly, the figure shows that

$$P(\chi^2_{21} \geq 12.7) \approx 92\%. \tag{45}$$

We shall see shortly that both the above values can be related to the experiment we studied in the last section (Table 6). But first a few general remarks about the problem we are going to tackle.

Let $x_1, x_2, \ldots, x_n$ be numbers obtained from n independent observations of an experiment, where the experimental conditions are identical from one experiment to another. This is the formulation in everyday language. We translate this into a mathematical model by assuming that there is a probability space (Ω, P) and n independent, identically distributed random variables $X_1, X_2, \ldots, X_n$ together with a point $\omega_0 \in \Omega$ such that $X_1(\omega_0) = x_1$, $X_2(\omega_0) = x_2, \ldots, X_n(\omega_0) = x_n$. In other words (see Exercise 17), $x_1, x_2, \ldots, x_n$ is a *sample*.

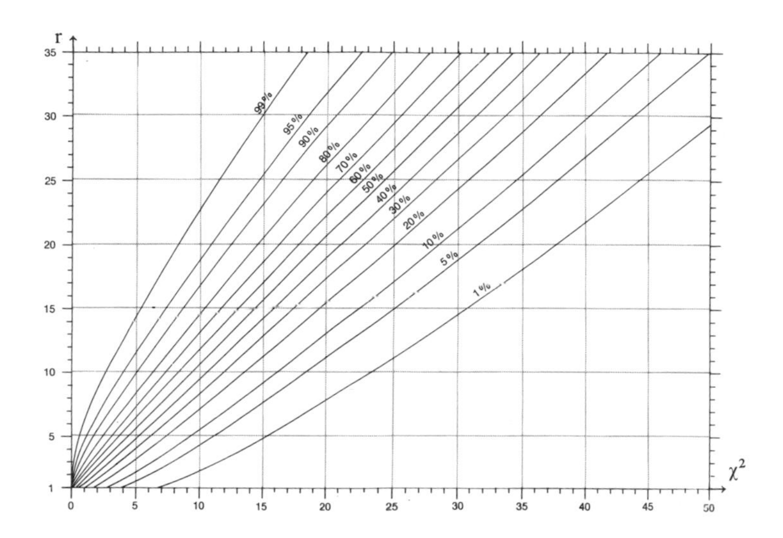

Fig. 11. Curves showing fractiles in χ^2-distributions.

Note that the model itself does not allow us to draw a single conclusion about the real situation we are interested in. It is the model in relation with the mutual interpretability that exists between probability theory and reality which allows us, in certain situations, to compute quantities (numbers, sets of numbers, or functions) that have an interpretation, making it natural to use them as a basis for statistical opinions.

We have seen examples of this earlier. The problem we shall treat now is somewhat more complicated. It gives a good opportunity to pick up an important aspect of statistical thinking.

In reality, the model suggested above may appear fairly absurd at first glance. We realise well enough what role ω_0 will play. It is to encode the sample we have obtained. But what about the other elements of Ω? Do they play any role at all? Indeed they do! The fundamental idea is that they encode *what could possibly have happened.* And we shall relate this to what actually *did* happen. In what follows we shall see how this can be done.

THESIS 26. *A statistician is just as interested in what could have happened as in what actually did happen.*

As described in §16, a sample determines a certain distribution, the empirical distribution. We denote the latter by G. Then G is a mapping of the type

$$I \mapsto G(I); \qquad I \subseteq \mathbf{R}.$$

For $I \subseteq \mathbf{R}$ (we use the notation I because I is usually an interval), $G(I)$ equals $1/n$ times the number of k for which $x_k \in I$. Thus the actual number of observations which lie in I is $G(I) \cdot n$.

Now suppose that we want to check the conjecture that the sample comes from some particular distribution we denote by G^*. The distribution G^* is, like G, a mapping of the type

$$I \mapsto G^*(I); \qquad I \subseteq \mathbf{R}.$$

Such a conjecture can arise in many different ways. G^* could be a distribution laid down in advance, e.g., a uniform distribution when the experiment we undertake is throwing a coin or dice, or G^* could be determined from provisional modelling, with one or more unknown parameters in the model having been estimated by means of the sample, as we did in the example in the previous section. In the latter situation, where parameters are estimated, we shall assume in the future that this is done by the maximum likelihood method.

The conjecture that a sample is from a distribution G^* can be expressed by the probability measure P. The conjecture amounts to saying that for each system $I_1, I_2, \ldots, I_n$ of subsets of $\mathbf{R}$,

$$(46) \quad P(X_1 \in I_1, X_2 \in I_2, \ldots, X_n \in I_n) = G^*(I_1)G^*(I_2) \cdot \cdots \cdot G^*(I_n)$$

Thus the conjecture gives a more precise specification of the model; in fact, as it is only distributions which interest us (cf. Theses 13, 14, and 15), one can say that (46) completely determines the model. We know that there is a certain freedom in the choice of probability space and random variables, but this freedom does not affect the later probabilistic and statistical considerations at all. One can remove the slight "mystery" which surrounds the partially indeterminate elements of the model by remarking that it is always possible to use $\Omega = \mathbf{R}^n$ and to use the coordinate mappings as random variables.

Let us elaborate on the last remark by looking at the case $n = 2$. Here we set

$$\Omega = \mathbf{R}^2 = \{(a_1, a_2) \mid a_1 \in \mathbf{R},\ a_2 \in \mathbf{R}\}$$

and define the random variables $X_1 : \Omega \to \mathbf{R}$ and $X_2 : \Omega \to \mathbf{R}$ by

$$X_1(a_1, a_2) = a_1, \qquad X_2(a_1, a_2) = a_2; \qquad (a_1, a_2) \in \Omega.$$

It remains to define P so that (46) is satisfied. For a product set, i.e., a set of the form

$$I_1 \times I_2 = \{(a_1, a_2) \mid a_1 \in I_1,\ a_2 \in I_2\},$$

we define P by

$$P(I_1 \times I_2) = G^*(I_1) \cdot G^*(I_2).$$

This determines the probability for all rectangles parallel to the axes and we have ensured that (46) holds for such sets. One can then use axioms for probability measures (see Exercise 3) to determine $P(A)$ gradually for more and more complicated events A. E.g., it is clear how P(A) can be defined for the event sketched in Figure 12.

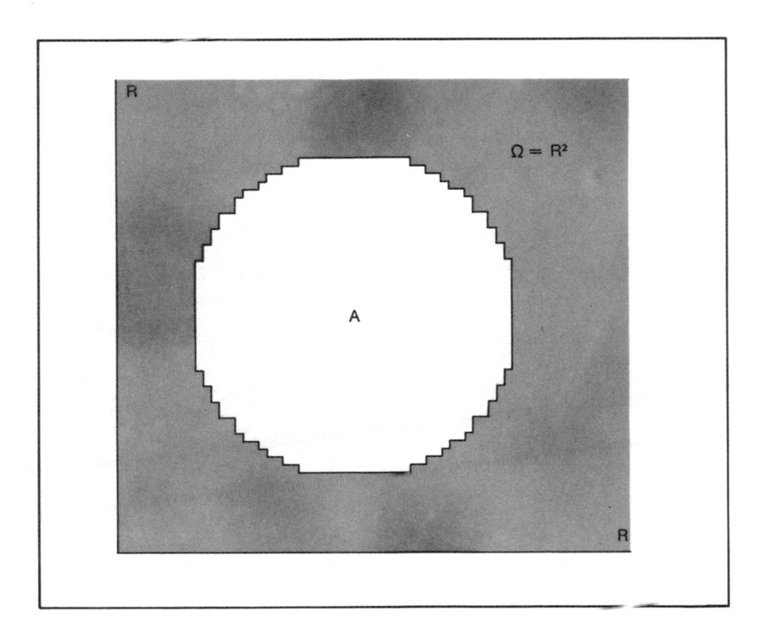

Fig. 12.

Think about it! A complicated theoretical analysis shows that if the process of determining $P(A)$ for more and more complex events is continued, it leads "at last" to a complete specification of the probability measure.

We turn back to the concrete problem of comparing G with G^*. We shall give a method of measuring the "distance" or, as we shall also say, the *discrepancy* between G and G^*. The first step is to choose a partition, here often called a *grouping*, of $\mathbf{R}$ into finitely many classes. We denote the grouping by $(I_k)_{k=1,2,\dots,\nu}$, and hence we have $\mathbf{R} = \cup_1^{\nu} I_k$ where $I_j \cap I_k = \emptyset$ for $j \neq k$. If some set $S \subseteq \mathbf{R}$ contains both the support of G and the support of G^*, then it is enough to ask that $(I_k)_{k=1,2,\dots,\nu}$ be a partition of S.

We now define the observed and expected magnitudes for the grouping by

$$N_k = n \cdot G(I_k), \qquad N_k^* = n \cdot G^*(I_k); \qquad k = 1, 2, \cdots, \nu. \tag{47}$$

In other words, N_k is the number of observations in I_k, whereas N_k^* is the corresponding expected number, assuming that the conjecture that the sample comes from the distribution G^* is correct.

We can now define the *distance* or *discrepancy* between G and G^* (relative to the grouping) as the number

$$d = \sum_{k=1}^{\nu} \frac{(N_k - N_k^*)^2}{N_k^*} = \sum_{k=1}^{\nu} \frac{N_k^2}{N_k^*} - n. \tag{48}$$

Let us look more closely at this quantity, and investigate, just qualitatively, the extent to which it satisfies the expectations we have a measure for the distance between two distributions.

The distance is defined as the sum of ν terms, and we can think of each individual one as measuring the divergence between the observed and expected values on the associated set I_k. It is clear that if N_k^* becomes very large we must tolerate larger divergence between N_k and N_k^*. It is precisely to compensate for this state of affairs that N_k^* appears in the denominator in (48). However, that has the effect that d is in danger of becoming unreasonably large when there is a k for which N_k^* is very small. The latter situation must therefore be avoided.

There is another important circumstance which should be pointed out. If we have used the sample $(x_1, x_2, \ldots, x_n)$ to estimate several unknown parameters it is clear that we must expect comparatively better agreement between G and G^*, which means that d will be comparatively small. There is nothing undesirable in this, it is just that to be able to consider G and G^* "close" together we have to impose stronger demands on d (d must be very small), the more parameters we have estimated with the help of the sample.

How large can one allow d to be and still uphold the conjecture that the sample is from the distribution G^*? We shall decide this by viewing d as a random variable. Certainly, d is defined as a definite number, but it is intuitively clear that the value taken by d is accidental. It depends on the sample $x_1, x_2, \ldots, x_n$. To formalise this we can use the model in which the sample is viewed as values of the random variables $X_1, X_2, \ldots, X_n$. In other words, when we think of d as a definite number, it is the discrepancy actually present, whereas the random variable approach indicates the discrepancy we could have had.

It is convenient to distinguish between two cases.

The easiest is that which assumes that G^* is a fixed distribution independent of the sample (no parameters have been estimated). The N_k^*'s are then fixed numbers. On the other hand, we can regard the N_k's as random variables, by defining

$$N_k(\omega) = \frac{1}{n} \cdot \text{ number of } i \leq n \text{ for which } X_i(\omega) \in I_k; \qquad \omega \in \Omega.$$

Thus the discrepancy d can be regarded as a random variable merely by replacing the N_k in (48) by $N_k(\omega)$. Since it is very important that we distinguish sharply between the two ideas, we use D to denote the random variable and, as before, d to denote the present value. We therefore have

$$(49) \qquad D(\omega) = \sum_{k=1}^{\nu} \frac{(N_k(\omega) - N_k^*)^2}{N_k^*}; \qquad \omega \in \Omega.$$

If a sample is used to estimate one or more parameters, the N_k^*'s can also be regarded as random variables. Substituting these in (48) as well gives the random variable corresponding to d. This random variable is likewise called D.

Thus in all cases we have defined a random variable D. We can use the latter to define a *significance level* SL:

$$SL = P(D \geq d).$$

Here $D \geq d$ is a short notation for the event consisting of the $\omega \in \Omega$ for which $D(\omega) \geq d$.

Suppose, e.g., that $SL = 20\%$. When G^* is a fixed distribution, this can be interpreted as follows: if a mass of samples are collected, all of size n, if all samples are from the distribution G^* and if they are all independent of each other, then for c. 20% of the samples the discrepancy will be as big as or bigger than, the discrepancy actually observed in the sample we took. Thus the discrepancy can hardly be said to be unacceptably large, since in 1 out of 5 cases we can expect to get a discrepancy at least that large.

If the significance level is 1% then we have observed a discrepancy so large that we can only expect to find another as large in 1 out of 100 cases. Then of course we would naturally reject the conjecture that G^* is the underlying distribution, and we would look for another model.

If we are in the situation where the sample is used to estimate one or more parameters, then we have a completely parallel interpretation of the meaning of the significance level. One just has to remember that G^* is not a fixed distribution (G^* is a random distribution!).

We can summarise the above considerations by introducing the following *test for goodness of fit*. Choose a suitable limit α for the significance level. If $SL < \alpha$, one rejects the conjecture that the underlying distribution is G^*, and if $SL \geq \alpha$ one maintains the conjecture (whether one can simply consider the conjecture to be confirmed on these grounds is still doubtful). The choice of α is a matter of taste and experience. If the material is somewhat uncertain and it is clear that there are many factors one has not taken into account in setting up the model, then one can defend the choice of a rather low value for α. A value of α around 5% is often reasonable.

It should perhaps be noted that if one finds, e.g., a significance level of 90% that does not mean that the conjecture is confirmed to a far higher degree than when one finds a value of, e.g., 25%.

Our method for testing goodness of fit is not very valuable if we cannot compute the significance level. We have the following deep result, formulated in the terms of the foregoing discussion:

THEOREM 4. *If the conjecture about the distribution from which the sample comes is correct, and if all the expected values N_k^*; $k = 1, 2, \ldots, \nu$ are large, then the significance level is given by the following approximate formula:*

$$SL \approx P(\chi^2_{\nu-s-1} \geq d), \tag{50}$$

where s denotes the number of parameters which have been estimated from the sample.

Because of the role played by the χ^2-distribution in this result, we speak of the χ^2-*test for goodness of fit* when using it.

It is difficult to say anything more precise about when one can allow oneself to use the theorem. However, we shall give some loose guidelines. Experience shows, regarding the magnitude of the expected values N_k^*, that one only has to make sure that they are all at least 5. Formula (50) is an approximation which becomes better as the size n of the sample increases. However, the approximation is so good that in practice one can apply the formula as a matter of course (it is only in the case where a very "irregular" distribution is under investigation that one should be on guard).

Note that the theorem works well in situations where the number of groups, ν, is very small (though of course ν must not be 1).

The theorem assumes, strictly speaking, that some quite definite conditions are satisfied. First, the grouping must be fixed in advance; thus it may not be determined from the sample. Second, any estimations of parameters must be based on the *grouped sample*, i.e. from the numbers κ_k; $k = 1, 2, \dots, \nu$, where $\kappa_k = \sum\{x_i \mid x_i \in I_k\}$. In addition, as already indicated, the estimates must be carried out by the maximum likelihood method.

The extra inaccuracy one introduces by ignoring these demands will usually be small. Therefore one will often arrange the statistical analysis without taking the demands into account, but so that the procedure and computation as a whole are simplified. A convenient way of proceeding is the following: first determine, on the basis of the *whole sample*, the distribution G^*. Then use G^* to determine a grouping so that the N_k^*'s are at least 5, say, and finally use the theorem to calculate the level of significance.

The convenience of this, only slightly inaccurate, procedure can be seen, e.g., in the case important to us, where the model G^* is a Poisson distribution, with parameter λ not known in advance. Here we are lucky because there is a very simple way to estimate λ, namely by the sample average (cf. Exercise 17). But this is not valid if we have made a grouping (unless the latter is "uniform"). One can of course easily carry out an estimation of λ based on a grouped sample, but there is no practical "ready-made" formula, and we are obliged to carry out a lot of calculation. Here there may be a need to develop a program so that the calculation can be done conveniently by machine. It is hoped that our discussion has been thorough enough to enable the interested reader to prepare a reasonable program himself.

We stick to the more convenient way of proceeding and for this purpose we have worked out Programs $P7$ and $P8$ (corresponding to the case where G^* is a Poisson distribution or modified Poisson distribution respectively). But one must bear in mind that too strong a grouping will only give limited accuracy.

Let us see what happens with the numerical data from the previous section (Table 6). The best Poisson approximation is described in Table 8. It is based on the sample used to estimate the parameter λ. A corresponding grouping $(I_k)_{k=1,2,\dots,\nu}$ together with the set of observed and expected numbers (rounded to the nearest integer) is given in Table 8.

Table 8 Observed and expected numbers according to the Poisson model

k	1	2	3	4	5	6	7	8	9	10	11	12	13	14	15	16	17	18	19	20	21	22	23	24	25	26	27	28	29
I_k	≤58	≤61	≤63	≤65	66	67	68	69	70	71	72	73	74	75	76	77	78	79	80	81	82	83	84	85	86	≤88	≤90	≤93	<∞
N_k	0	4	5	6	4	6	9	11	8	11	17	13	19	19	14	14	16	10	8	10	11	9	5	5	3	3	5	0	4
N_k^*	5	7	8	11	7	7	8	9	10	10	11	11	11	11	11	11	11	10	10	9	8	8	7	6	5	9	6	6	5

Table 9 Observed and expected numbers according to the modified Poisson model

k	1	2	3	4	5	6	7	8	9	10	11	12	13	14	15	16	17	18	19	20	21	22	23	24
I_k	≤62	≤64	≤66	67	68	69	70	71	72	73	74	75	76	77	78	79	80	81	82	83	84	85	≤87	<∞
N_k	5	6	8	6	9	11	8	11	17	13	19	19	14	14	16	10	8	10	11	9	5	5	3	12
N_k^*	5	5	9	6	8	9	10	12	13	14	14	15	15	15	14	13	12	11	10	8	7	6	8	10

The sets I_k are given as subsets of $\{0, 1, 2, \dots\}$. We have chosen the I_k's so that the corresponding expected number becomes at least 5. We say that we have made a *grouping with group limit* 5 . The table was worked out with the help of $P7$, which also gives the discrepancy. The latter is $d = 42.98$. Since the number of degrees of freedom is $29 - 1 - 1 = 27$, we see from Theorem 4 together with Figure 11 that the significance level is *c.* 3%, cf. (44). Thus there are good grounds for rejecting the pure Poisson model.

For the model with the modified Poisson distribution, which requires estimation of λ and h (see §16), we find the data given in Table 9 for the observed and expected numbers. Here $d = 12.73$ and, since the number of degrees of freedom is $24 - 2 - 1 = 21$, Theorem 4 and Figure 11 give a significance level *c.* 92%, cf. (45). Programs P3 and P8 were used in calculating these data.

18. The Historical Perspective

Two main topics will be discussed. One is the development of the mathematics we have used; the other is the development of the branch of physics we have been concerned with.

As far as the mathematics is concerned, probability theory has been our essential tool. This branch of mathematics had its origin in various games such as dice, cards, etc. In particular, one wished to be able to compute the odds for different game situations.

The first sound basis for probability theory was established in 1654 in the famous correspondence between Fermat and Pascal. Because of an important preceding work of Cardano, one speaks of the *Cardano-Fermat-Pascal probability concept.* It regards a probability as "The number of favourable cases divided by the number of possible cases."

In the period that followed, most work was with combinatorial problems. The binomial distribution (with success probability $\frac{1}{2}$) was one of the most important objects of study. But there were difficulties with the Cardano-Fermat-Pascal probability concept, as many realised. In his book *Ars Conjectandi*, published posthumously in 1713, Jakob Bernoulli wrote in detail about situations where probabilities originating in nature or human behaviour, that it is meaningless to try and count the number of "favourable cases." Here Bernoulli pointed to the possibility that what one could not determine *a priori*, one could determine *a posteriori.* It is to Bernoulli's great credit that he succeeded in formalising this view in a precisely formulated mathematical theorem, the *law of large numbers.* Bernoulli proved his theorem for the binomial distribution; it

Fig. 13. *Abraham de Moivre (1667-1754) was born and educated in France. Like many other protestants, he took flight to England after the revocation of the Edict of Nantes in 1685. In England de Moivre made a living by giving private lessons in mathematics. He studied Newton's "Principia" by tearing pages out of the book to read as he went from one pupil to another. He dedicated the first edition of his famous book "The Doctrine of Chances" to the ageing Newton. Even though Newton and the scientific world ranked de Moivre's ability and judgement highly - he was appointed to judge the priority dispute between Leibniz and Newton, for example – he continued to survive only by giving private lessons.*

has since been extended to a completely general result.

In 1733 de Moivre refined the investigations of the binomial distribution and proved the *central limit theorem.* In de Moivre's theorem the success probability was $\frac{1}{2}$. The result was extended to an arbitrary binomial distribution in 1812 by Laplace.

Let us turn to Poisson's contribution. In 1837 Poisson published a book *Recherches sur la probabilié des jugements en matière criminelle et en matière civile.* In it he made a weighty contribution to the debate over the appropiateness of using probability theory (and also other mathematical methods) in the area of social sciences.

Politically conservative circles and also philosophers such as Auguste Comte did not think that such a rationalisation was possible or desirable.

With energy Poisson engaged himself in the debate and defended the universal character of probability theory. In a way, the discussion was in line with the viewpoint expressed by Bernoulli 100 years earlier, and Poisson also quoted Bernoulli's main result as his central point. Moreover, it was Poisson who introduced the term "law of large numbers."

In studying limit theorems for the binomial distribution, Poisson also treated a problem that in a certain sense can be said to have escaped Laplace's notice, namely the case where there is great "asymmetry" between the probabilities of success and failure, while the product of the number parameter and the success probability remains constant under passage to the limit. In other words, Poisson derived the distribution that today bears his name in proving Theorem 3 of §14.

Poisson's contribution to probability theory does not bear comparison with Bernoulli's, de Moivre's or Laplace's. His ideas were not creative to the same degree. Rather, he consolidated the existing theory. One can even question whether he was the first to introduce the Poisson distribution and its relation to the binomial distribution, since de Moivre had also touched on it, though in such an indirect form that it would be an abuse of hindsight to give de Moivre full credit.

Poisson's work was not appreciated by his contemporaries, who considered it to be an echo of Laplace's – it was scarcely appreciated even for its merits as an insightful and well written popularisation. An exception to this was the Russian school, which was being founded at the time by Chebyshev.

In reality, with Laplace's work probability theory had reached the point where it was difficult to go further. There was a need for a new foundation. The Cardano-Fermat-Pascal probability concept had not been able to meet many of the challenges to probability theory,

Fig. 14.*Siméon-Denis Poisson (1781-1840) studied mathematics under Laplace and Lagrange at the École Polytechnique, where he became professor himself in 1808. He was particularly interested in mathematical physics. Towards the end of his life he published two long works, one the probabilistic work mentioned in the text, the other "Recherches sur le mouvement des projectiles dans l'air." In the latter we see an example of the way in which Poisson used the results of others without scruple: an important improvement in the description was his taking account of the Coriolis effect, but without mentioning Coriolis, in spite of the fact that Poisson himself had judged Coriolis' doctoral work. It is therefore not surprising that Poisson got a bad reputation, and that it was long after his death before Poisson's contributions, both purely scientific and administrative, appeared in a more balanced and favourable light.*

Fig. 15. *Andrei Nikolayevich Kolmogorov (born 1903) is perhaps the century's most influential Soviet mathematician. His interests extend widely and include, besides purely mathematical fields, such as probability theory, logic and information theory, the fields of pedagogy and old Russian art. He has to an outstanding degree been capable of setting a fashion and had many followers; it is said that his students undertook hard physical training in order to be able to keep up on the long walking and ski excursions on which Kolmogorov liked to discuss mathematical subjects. Added note to the English edition: Kolmogorov died on October 20, 1987. For an informed obituary see The Times, London from October 26, 1987.*

e.g. from population statistics and from actuarial science.

The best weapon one had, and this was just what Poisson used, was appeal to the law of large numbers. This law is of enormous importance. The view of probability theory it expresses (cf. Thesis 12) underlies many considerations, particularly those of a more statistical character, as in this book. The theoretical possibility the theorem points to, that of determining probabilities (and mean values), gives reason for interpretations which serve as an important guide. But it is not appropriate, in the opinion of most probability theorists (including the author), to seek a foundation for all of probability theory that is built directly on the law of large numbers. This *can* be done today, but even though this is mathematically exciting, one has to say that such a foundation does not make possible the desired fruitful interaction between probability theory and reality.

The firm foundation, which the probability theory had been missing, was first established in Kolmogorov's book *Grundbegriffe der Wahrscheinlichkeitsrechnung* of 1933. It should be mentioned that Kolmogorov built upon the ideas of Borel, among others, and also those of the Russian school of probability theorists (Chebyshev, Markov, Lyapunov).

The starting point of Kolmogorov's theory is the concept of a *probability space* (see Exercise 3). A main result of the theory, the so-called *consistency theorem*, involves the construction of "new probability spaces from old." The result can be interpreted as saying that every stochastic phenomenon that is merely specified in the form of certain distributions, has a model in Kolmogorov's theory. All that is required is that the distributions satisfy some "consistency conditions" which in practice are satisfied automatically. This theorem shows that a long series of stochastic phenomena which had been studied – but with a certain hesitancy since one could not indicate a precise model – could now be treated on a secure basis. Thus it was with Kolmogorov's model that one was first able to give a satisfactory description of, e.g., Poisson processes.

Kolmogorov's work, in the *Grundbegriffe* and works which followed, concerning stochastic processes especially, has been of decisive significance in the breakthrough of probability theory to its position as a respected, modern science with wide applicability.

Even though a firm foundation was not available for a long time, the development nevertheless did not stand still. Here we shall be content to comment on some investigations directly related to the subjects we have worked with in this book.

Fig. 16. *Ladislaus von Bortkiewicz (1868-1931) was a statistician, economist and actuary. Born in St. Petersburg (now Leningrad), where he studied jurisprudence, Bortkiewicz took his doctorate in Göttingen under Wilhelm Lexis (see Exercise 33), and worked for a time in the Russian Ministery of Transport.*

From 1901 he was professor of political science at Friedrich-Wilhelm-University in Berlin. Bortkiewicz's work is still rather underestimated, which may be partly due to his style and partly because he stood in the shadow of the English statistical school.

Fig. 17. *William Sealy Gosset (1876-1937), after studies at Oxford, was employed by the Guinness brewery in Dublin. He became interested in statistics, and published a series of works under the pseudonym "Student". He had great insight into technical problems of brewing, and there was always a close connection between his statistical research and the practical work he was engaged in. Gosset had numerous interests, among them fruit-growing (especially the cultivation of pears!), boat-building and many sports. The last years of his life he was tormented by the after-effects of a serious motorcycle accident.*

In 1898 Bortkiewicz showed the scope of the Poisson approximation to the binomial distribution in a long series of examples, among them the example of the fatal horse kicks (Table 3). Bortkiewicz often returned to the Poisson distribution, particularly to statistical investigations of it. Of special interest to us is the book he published in 1913, *Die radioaktive Strahlung als Gegenstand wahrscheinlichkeitstheoretischer Untersuchungen*. This treats, among other things, the Rutherford-Geiger experiment, the results of which had been published only two years earlier. It has surprised and impressed me to see that Bortkiewicz, with some long calculations, was in a position to estimate the dead time in the Rutherford-Geiger experiment (see Exercise 31). In fact, it is true of the work of Bortkiewicz and many others who will be mentioned in this section that it exhibits a hearty appetite for complicated calculations based on the mathematical analysis.

Gosset, whom we have mentioned in §15, worked as a statistician at the Guinness brewery in Dublin. As is well known, brewing is a delicate affair. Many parts of the production process require working to close tolerances. Consequently, Gosset had good reason to seek refined statistical methods. Under the pseudonym "Student," he published a series of recognised works originating from concrete problems of brewing. Here we point out his first work, "On the error of counting with a hæmocytometer," from 1907. The numerical material of Table 5 comes from the latter work. An important point is that the range of applications of the Poisson distribution was extended as indicated in §15. It should also be mentioned that the (unusually?) well-formulated Thesis 25 is taken directly from one of Gosset's statistical works ([14], p. 149).

In the sources we have been concerned with until now, the Poisson distribution appears as an isolated distribution. One can ask when the Poisson processes came to life. Of course one can be very critical and answer that stochastic processes did not exist at all before Kolmogorov's work. But that would be pedantic. The characteristic view of processes is that one incorporates the time parameter in an essential way, and the question is: when did this aspect first show up?

We turn to the study of telephone traffic. Here the development in time plays a prominent role.

Surely, only few readers are aware that in the field of telephone communication Denmark is a pioneering country. It was in this country that probability theory was first applied in a significant way to the field of telephony. This occurred in 1907 and 1908 in two essays by the administrative director of the Kjøbenhavns Telefon Aktieselskab, F. Johannsen.

In 1908 the company was quite unique and far-sighted in establishing a physical-technical laboratory for scientific development. The mathematician A. K. Erlang was appointed leader. Soon afterwards, in 1909, Erlang published his first work on the subject of telephone technology. He showed, among other things, that the number of calls to a central exchange can be described, under natural idealised assumptions, as a Poisson process. This is without doubt one of the first works in which a Poisson process appears, though admittedly in a somewhat rudimentary form. Whether it is actually the first, I do not know for certain.[14]

In Erlang's further work he introduced a series of distributions for the study of processes that are more complicated than the Poisson process, in as much as one must take account not only of the arrival of events (telephone calls) but also of the cessation of events – after all, most conversations come to an end. Also, one must take the limited number of subscribers into account.

Erlang's works were nearly all connected with problems of dimensioning in telephone technology and were of great practical significance. They quickly spread beyond the borders of his country. A work of 1917, in which Erlang calculated the probability of a call being turned down by the central exchange because of conversations already in progress, was particularly important. In order to carry out such a calculation Erlang introduced a principle of "statistical equilibrium." This principle anticipated later developments and plays an important role to this day.

Erlang's work lives on, both in practice and as a basis for mathematical and technological-scientific investigations. A sign of the appreciation his contributions have earned is the fact that the international unit for telephone traffic bears his name.

The telephone companies in Denmark have continued the good tradition by making the most of mathematical research talent in the telephone service. There are, of course, other examples of fruitful exchange between industry and mathematics. But there are many still untried possibilities where farsightedness from industry and interest from mathematicians could be mutually gratifying. To conclude these remarks, may I mention the prosaic

THESIS 27. *Mathematics is worthwhile.*

---

[14]Since the Danish edition was published, the author has learned that the Swedish actuary Filip Lundberg published a work in 1903 which also introduced Poisson processes (and more general processes). The work is the forerunner of the so-called collective risk theory which Lundberg developed later.

Fig. 18. *Agner Krarup Erlang (1878–1929) was born in Lønborg at Tarm, Denmark. Many in this country (Denmark) know him best for his tables, which were often used in schools before pocket calculators made their appearance. His most important contribution, however, was as a poineer in the field of telephone communication. He worked in the Kjøbenhavns Telefon Aktieselskab (K.T.A.S.) from 1909 until his death. Erlang's personality bore the stamp of a philosophical way of thinking and a religious disposition.*

The statistical methods we have used were introduced by the powerful English school. The maximum likelihood method was developed in 1912 and later by the famous statistician R.A. Fisher; however, a special case of the method goes back to Gauss. It was K. Pearson, who in 1900 introduced the χ^2-test, but an important later development, corresponding to the case where parameters are estimated, is due to Fisher (1922 and 1924). A long series of works which followed, down to our own time, further developed the theory of this important test.

We have left until last the developments directly related to the main physical subject of our book, radioactivity.

Radioactivity was discovered in 1896, more or less by accident, by Henri Becquerel. The first to investigate the phenomenon more closely were Marie and Pierre Curie. One cannot help being moved by their persistent and unselfish contributions, especially their work in chemistry to isolate new radioactive materials (polonium and radium) and to exploit the possibility of using radioactivity in the service of medicine.

It was Marie Curie who introduced the term "radioactivity." Together with Becquerel, the husband and wife Curie received the 1903 Nobel prize for their study of radioactivity. In view of the revolution in human affairs brought on by the discovery of radioactivity, it is only fair to cite the following from Pierre Curie's speech on receiving the Nobel prize:

We might still consider that in criminal hands radium might become very dangerous; and here we must ask ourselves if mankind can benefit by knowing the secrets of nature, if man is mature enough to take advantage of them, or if this knowledge will not be harmful to the world .... I am among those who believe that humanity will derive more good than evil from new discoveries.

In spite of the contributions made by others, one has to say that it was Ernest Rutherford who became the leading explorer of the more fundamental side of radioactivity. He found different forms of radiation and introduced the terms "alpha" and "beta" radiation. Together with the chemist Soddy he arrived in 1902-03 at the "alchemical" view that

Radioactivity is at once an atomic phenomenon and the accompaniment of a chemical change in which new kinds of matter are produced.

This view led to an extensive project of ordering all radioactive elements into decay series. Rutherford and Soddy introduced the term half-life and realised that apparently constant activity could be explained by

Fig. 19. *Geiger (left) and Rutherford in the laboratory at Manchester, c. 1910.*

Hans Wilhelm Geiger (1882-1945) was a German physicist. He spent the years 1906-1912 in Manchester. In 1908 he developed, together with Rutherford, the forerunner of what we now call the Geiger counter or the Geiger-Müller counter. The final construction was carried out in joint work with Walther Müller, shortly after Geiger became professor in Kiel in 1925. The new counter was the basis of a series of detailed investigations of cosmic rays undertaken by Geiger.

Ernest Rutherford (1871-1937) was born in New Zealand. With his energy, enthusiasm and ingenuity, he demonstrated his ability early. Independently of Marconi, he built a detector for radio signals. In 1895 he came to England, where he worked under J.J. Thomson in Cambridge. In 1898 he became professor of physics at McGill University in Montreal; in 1907 he came to Manchester and in 1919 succeeded to Thomson's chair in Cambridge. Rutherford laid the foundations for the development of nuclear physics with his exploration of radioactivity, his discovery of alpha-particles and his development of an atomic model. He received the Nobel prize for chemistry in 1908.

a half-life that was long in comparison with a human lifetime. The understanding of the quantitative properties of radioactive processes that was gained in this way was enormously important, and led to a correction of the Curies' conception, among other things.

A long series of investigations followed. Here we merely mention that around 1905 it first became possible, by age measurements based on radioactive series, to get an idea of the age of the earth (Boltwood and Rutherford). Rutherford determined the age of some uranium minerals to be around 500 million years. This was sensational, since Lord Kelvin, only about 10 years earlier, had maintained with great confidence that, according to complicated mathematical calculations based on different methods, the age of the earth was between 20 and 40 million years. It is interesting that radioactivity could also give an explanation of why Lord Kelvin's result was wrong. He had based his calculations on a "cold earth," but Rutherford was able to show that the presence of uranium accounted for quite significant heat production.[15] Geologists have long since accepted the radioactive dating method, and now give the age of the earth as approximately 4.6 billion years.

Rutherford, who until this time had worked at McGill University in Canada, moved to Manchester in England in 1907. Here he began his joint work with Hans Geiger, among others.

Rutherford's preferred objects of study were the α-particles he himself had discovered. He was of the opinion that they contained the germ of deeper understanding of the structure of matter.

Geiger and Rutherford developed various methods of observation that allowed a detailed study of α-radiation. The first one made possible an exact count of the number of α-particles emitted per second from a radium source. It was also at this time, in 1908, that they succeeded in directly confirming Rutherford's conjecture that α-particles are doubly ionised helium atoms.

We now come to the time of the Rutherford-Geiger experiment, 1910, which we have chosen as the main example in this book (see §10). At this stage the famous scattering experiment of Geiger and Marsden had been carried out and Rutherford had certainly gained a clear feeling for its consequences, in the form of a revision of the existing atomic model. But it was only in 1911 that he first published his famous model of the

[15]In this connection one can find, in Faure's book [9], the following quotation from T. C. Chamberlin: *The fascinating impressiveness of rigorous mathematical analysis, with its atmosphere of precision and elegance, should not blind us to the defects of the premises which condition the whole process.*

atom, whose central element he conceived to be an atomic nucleus which was a very small part of the atom itself.

In the introduction to the article, Rutherford and Geiger wrote very clearly about the background and purpose of their work:

In counting the α particles emitted from radioactive substances either by the scintillation or electric method, it is observed that, while the average number of particles from a steady source is nearly constant, when a large number is counted, the number appearing in a given short interval is subject to wide fluctuations. These variations are especially noticeable when only a few scintillations appear per minute. For example, during a considerable interval it may happen that no α particle appears; then follows a group of α particles in rapid succession; then an occasional α particle, and so on. It is of importance to settle whether these variations in distribution are in agreement with the laws of probability, i.e. whether the distribution of α particles on an average is that to be anticipated if the α particles are expelled at random both in regard to space and time. It might be conceived, for example, that the emission of an α particle might precipitate the disintegration of neighbouring atoms, and so lead to a distribution of α particles at variance with the simple probability law.

After a careful description of the experiment and an analysis of the results the authors could write:

We may consequently conclude that the distribution of α particles in time is in agreement with the laws of probability and that the α particles are emitted at random. As far as the experiments have gone, there is no evidence that the variation in number of α particles from interval to interval is greater than would be expected in a random distribution.

With this background we can assert:

THESIS 28. *Rutherford and Geiger were the first to recognise a phenomenon as being truly spontaneous, stochastic in principle.*

What is important here is the phrase "in principle." Earlier works based on probabilistic considerations, e.g., Boltzmann's statistical mechanics, really rested on a deterministic foundation, Newton's mechanics (cf. Thesis 8).

But we can maintain, with great confidence, that at least until Rutherford and Geiger nobody could adopt a genuinely stochastic foundation. However, despite some convincing passages we could cite from their work, it is doubtful whether Rutherford and Geiger were completely clear about this new element. They conclude their paper as follows:

Apart from their bearing on radioactive problems, these results are of interest as an example of a method of testing the laws of probability by observing the variations in quantities involved in a spontaneous material process.

What a mistake! There is absolutely no basis for testing the laws of probability theory by observing nature. The laws have their internal mathematical life independent of the real world (see also the discussion in §11). Nevertheless one can investigate, like Rutherford and Geiger, whether a natural phenomenon can be explained by bringing in a mathematical model. If that succeeds, it will always involve an interpretation of the terms that appear in the model. Interpretation is a matter that lies mainly, though not exclusively, outside the domain of mathematics. Interpretations will normally be capable of application to situations other than the actual one, and this means important opportunities for testing *interpretations* – not mathematics! – by making further experiments.

The fact is that Rutherford and Geiger did not have sufficient mastery of the parts of probability theory that come into play here. As further evidence for this there is the fact that they asked a colleague at Manchester to work out an appendix on probability theory. In this appendix the Poisson distribution is derived – without any reference to previous works, incidentally.

Even though Rutherford and Geiger looked at the probability theoretic aspect, they did not recognise the break with classical notions that this implied – and who could blame them for that?

As mentioned before, Rutherford's model of the atom was published in 1911. Only a few saw its implications and possibilities. One of them was Niels Bohr, who first met Rutherford in 1911 and later spent long periods in Manchester. Perhaps one can say that whereas Rutherford's ability in theoretical physics was closely connected with his ability to devise and carry out experiments, Bohr's ability was of a largely independent character, with strong philosophical overtones, which allowed him to combine different views in unorthodox ways.

It was Bohr's success in 1913, with the application of quantum concepts to the theory of the hydrogen atom, that clearly showed the significance of Rutherford's model of the atom.

After 1913, rapid development took place. Around 1925 the ideas of Bohr, in particular, crystallised into a great connected whole, the "old

Fig. 20. "Does God play dice?" – *Einstein and Bohr at the Solvay congress in Brussels in 1930.*

Albert Einstein (1879–1955) was born in Ulm in Germany. Since Einstein, after completing his studies, had difficulty in obtaining work at an institute of higher education, he accepted a job as a patent clerk in Berne. In fact, this suited Einstein well. He found the rather uncomplicated work with patent applications a useful distraction from scientific investigations. It was during this period that he laid the foundations for his theory of relativity. He left the patent office in 1909 and worked from then on in scientific positions, for a long time in Berlin (1914–1933), and thereafter at Princeton in the U.S.A.

Niels Bohr (1885–1962) must be regarded as Denmark's most prominent physicist of all time. Bohr was copiously productive and carried on an extensive correspondence. His complete works, which are in the course of publication at present, are expected to fill 11 volumes. As an indication of the great responsibility Bohr felt as a pioneer of modern physics, we mention that in 1943, and again in 1950, he sought, in a letter to the UN, to get the major powers to show greater openness regarding nuclear weapons, as a means of hindering their use and securing peace. From 1920 onwards, Bohr led the newly established Institute of theoretical Physics at Copenhagen University and made it an international centre for nuclear research. Today the Institute bears Bohr's name.

quantum theory." The latter broke with classical concepts and in their place set up a series of higher principles for the interpretation and explanation of atomic processes. The principles contained both elements of a wholly new character and elements which gave a connection with traditional concepts.

The theory was, however, not as clear and comprehensive as classical Newtonian mechanics, and since the latter was no longer adequate, it was necessary to find a replacement. Such a replacement, in the form of the so-called *quantum mechanics*, was established in 1925 and 1926 with contributions from Heisenberg, Born, Jordan, Dirac and Schrödinger in particular.

One of the central aspects of quantum mechanics is the fundamentally stochastic conception of atomic phenomena. The latter represents such an emphatic break with traditional ideas that it has provoked much opposition. The germ of the stochastic view lies hidden in Planck's famous work of 1900, where the energy quantum was discovered, but Planck's discovery in no way had the stochastic view as a clear logical consequence. One has to say that Bohr, especially, was responsible for the latter viewpoint coming to stand so prominently in the foreground.

Among the opponents of the new theory we also find Einstein. Half teasingly, he could ask its supporters whether they really thought "that God plays dice." It is paradoxical that Einstein himself was one of the first, in a famous work of 1917 on radioactive equilibrium, to choose a stochastic conception as the starting point for his investigations. An essential feature which Einstein wished to treat was the spontaneous emission of radiation, and he himself drew attention to the fundamental character of the stochastic description, referring to the analogy with the description of radioactive phenomena, among other things. It appears, from some of his concluding remarks, that Einstein considered the stochastic framework to be a weakness in his work, a kind of bandaid solution which he hoped to be able to replace later by a more reasonable starting point.

Numerous attempts were made to show that quantum mechanics was incomplete. Often this was done by cunningly devised "paradoxical" thought experiments. But each time this happened, it turned out that the new theory was able to cover the situation. Einstein and Bohr were the main figures in such discussions. I refer the reader to the interesting account given by Bohr in his book [2] (pp. 45-82)[16]. Einstein's attitude

[16] A conversation with Margrethe Bohr, the widow of Niels Bohr, confirmed the impression one gets, in reading the latter account, that it was a great personal

can be illustrated further by the following statement:

I reject the basic idea of contemporary statistical quantum theory, insofar as I do not believe that this fundamental concept will prove a useful basis for the whole of physics .... I am, in fact, firmly convinced that the essentially statistical character of contemporary quantum theory is solely to be ascribed to the fact that this theory operates with an incomplete description of physical systems.

One should not reject the theoretical possibility, however distant it may be from actuality, that a much more detailed knowledge of the physical microcosm than we have today, will lead to a new physics resting on a different, perhaps fully deterministic foundation. The physicists who, in the early days of quantum mechanics, declared themselves sceptical, could not point to any such new possibility. The real reason for their opposition was an aversion to the idea of raising stochastic considerations to the status of a higher guiding principle. The opposition to this idea was not based on scientifically well-founded reasoning, as much as on a common human attitude of a philosophical-aesthetic character.[17]

One can see the breakthrough of quantum mechanics as a farewell to the clarity that had been gained through the long tradition since Newton's time, a clarity with a sound theoretical basis and, at the same time, with a strong leaning on common human experience related to observed behaviour of (familiar) material objects. It was the facts brought to light by experiments which compelled physicists to forsake their habitual lucidity. Whether new discoveries or new points of view one day will lead to the lost apprehensibility being regained is doubtful, and one has to say that developments in recent times indicate that a lucidity based, either directly or by analogy, on experiences that we can encompass with our senses is irretrievably lost.

---

disappointment to Bohr that he never succeeded in getting Einstein to enter the new quantum mechanics wholeheartedly

[17]It is interesting to draw a parallel with the attitude of an earlier time, when there was refusal to theorise about chance. The reasons for this were based on philosophical and religious considerations. Thus one can maintain that to a great extent it is because of Aristotle's dissociation, that probability theory developed comparatively so late.

Exercises

Exercise 1. If A and B are independent events, so too are each of the following events: A^c and B, A and B^c and A^c and B^c (little "c" denotes complement).

If A, B and C are independent events, so too are A^c, B^c and C^c.

Generalise these results if you feel like it!

Exercise 2. Make two throws of a die. Let A, B and C be the events

$$\begin{cases} \text{A: odd number on the first throw,} \\ \text{B: odd number on the second throw,} \\ \text{C: odd sum.} \end{cases}$$

Show that the pairs A and B, A and C, and B and C are independent, but that A, B and C are not independent.

Exercise 3. *A probability space* is a pair (Ω, P), where Ω (the *event space*) is a set and P (the *probability function*) is a function which to each *event* A, i.e. to each subset $A \subseteq \Omega$, assigns a number $P(A)$, in such a way that the following axioms are satisfied:[18]

(i) $0 \leq P(A) \leq 1 \qquad \forall A \in \Omega$

(ii) $P(\Omega) = 1$

[18]For the reader who knows the more complicated definition from advanced probability theory, I shall defend the simplification by reference to Solovay's result which implies, among other things, that there are models of set theory in which each subset of $\mathbf{R}$ is Lebesgue measurable, and universally measurable at that.

(iii) $A \cap B = \emptyset \implies P(A \cup B) = P(A) + P(B)$
(iv) $A_1 \subseteq A_2 \subseteq \cdots \implies P(\bigcup_1^\infty A_n) = \lim_{n\to\infty} P(A_n)$
(v) $A_1 \supseteq A_2 \supseteq \cdots \implies P(\bigcap_1^\infty A_n) = \lim_{n\to\infty} P(A_n)$

Perhaps the reader has not previously seen (iv) and (v) included in the definition of a probability space. They are important axioms nevertheless (though superfluous for finite event spaces). Prove the following:

(a) If $A_1, A_2, \ldots$ is a sequence of pairwise disjoint events, i.e., if $A_i \bigcap A_j = \emptyset$, $\forall i \neq j$, then for each n,

$$P(A_1 \cup A_2 \cup \cdots \cup A_n) = P(A_1) + P(A_2) + \cdots + P(A_n).$$

The reader who has been through Exercise 19 may also prove that

$$P(A_1 \cup A_2 \cup \cdots) = P(A_1) + P(A_2) + \ldots .$$

(b) In the axioms (i) - (v) one can do without either axiom (iv) or (v), since there is the equivalence:

$$\text{(i),(ii),(iii),(iv)} \iff \text{(i),(ii),(iii),(v)}.$$

Exercise 4. Suppose that the probability space (Ω, P) serves as a model for an infinite sequence of throws of a die, i.e., there is a sequence $X_1, X_2, \ldots$ of independent random variables defined on Ω such that

$$P(X_n = 1) = P(X_n = 2) = \cdots = P(X_n = 6) = 1/6 \qquad \text{for each } n.$$

Prove that the probability of such an infinite sequence of throws not including a single 6 is 0.

Hint: Use (v) from Exercise 3.

Let V denote the waiting time for the first 6 ($V = n$ means that the $1^{\text{st}}, 2^{\text{nd}}, \ldots, (n-1)^{\text{th}}$ throws are not 6's, but the nth throw is a 6). Prove that $P(V < \infty) = 1$ and determine the distribution of V.

Remark: The number of failures before success, $V - 1$ has a so-called *geometric distribution* with *parameter* $p = 1/6$.

Exercise 5. Suppose again that (Ω, P) gives a model for an infinite sequence of throws of a die (see Exercise 4). A_n denotes the event "6 at the nth throw." Describe in words the event

$$A = \bigcap_{n=1}^{\infty} \bigcup_{m=n}^{\infty} A_m$$

and compute $P(A)$.

Exercise 6. Prove that the distribution function F of a random variable satisfies the following conditions:

$$x_1 \leq x_2 \implies F(x_1) \leq F(x_2)$$

$$\lim_{x \to -\infty} F(x) = 0, \qquad \lim_{x \to \infty} F(x) = 1$$

$$F \text{ is continuous} \qquad \iff \qquad P(X = x) = 0 \quad \forall x \in \mathbf{R}$$

X is said to have a *continuous distribution* when its distribution function is continuous. Give examples of continuous distributions.

Exercise 7. Let X be a random variable. The *support* of X's distribution is defined to be the set of $x \in \mathbf{R}$ for which $P(x - \varepsilon < X < x + \varepsilon) > 0$ for each $\varepsilon > 0$.

Prove that the support is a *closed subset* of $\mathbf{R}$, i.e., for each convergent sequence of points in the support the limit point also belongs to the support (loosely speaking: it is impossible to get outside the support by passing to the limit).

Find the support in each of the following cases:

X is the number of points scored in a throw of a die.
X has Poisson distribution with parameter λ.
X has exponential distribution with parameter λ.

Exercise 8. Let X be a random variable. Let F denote its distribution function and let D be the support of the distribution. Suppose that the support is an interval (this will be an interval of the type $(-\infty, \infty)$, $(-\infty, b]$, $[a, \infty)$, or $[a, b]$).

Let x be an arbitrary point of $\mathbf{R}$ that is not an endpoint of the interval D. We say that the *density* of the distribution at x is α when, for a small interval I containing x,

$$P(X \in I) \approx \alpha \cdot \text{ length of } I.$$

More precisely, for each sequence of intervals $([y_n, z_n])_{n \geq 1}$ such that

$$y_n \leq x \leq z_n \qquad \text{for all } n$$

and

$$\lim_{n \to \infty} (z_n - y_n) = 0,$$

we demand that

$$\lim_{n\to\infty} \frac{P(X \in [y_n, z_n])}{z_n - y_n} = \alpha.$$

Prove that the density exists at x if and only if F is differentiable at x; in this case the density at x is simply equal to $F'(x)$.

If the density exists for each x that is not an endpoint of D, and if there is probability 0 of the random variable taking a value that is an endpoint of D, then we say that the distribution has a *density function,* and the latter is the function that associates to x the density at x. The density function is often denoted by f, and hence $f = F'$. Since the density function is 0 outside the support (think about it!), the density function is often given merely by specifying it on the interior of its support, i.e., on the support minus its endpoints.

Prove that each distribution that has a density function is continuous.

Prove that the distribution of a random variable that has a density function is completely determined by the density function.

Show that each exponential distribution has a density function, and determine the latter.

The *uniform distribution* over the interval $[a, b]$ is defined to be the distribution with density function

$$f(x) = \frac{1}{b-a}; \qquad a < x < b.$$

Draw a graph of the distribution function. Describe situations that can adequately be modeled by this distribution.

Exercise 9. In this and the next four exercises we shall look more closely at the mean value of random variables. We shall only consider non-negative random variables. This is a technical simplification, and in any case all the random variables we have been concerned with in the main text are of this type.

Let $(\Omega$,P) be a probability space and let $X \geq 0$ be a random variable defined on Ω.

Suppose that X is discrete. Then there is (see p.14) a countable set $A \subseteq [0, \infty[$ such that $P(X = a) > 0$ for all $a \in A$ and $P(X \in A) = 1$. We call the elements of A the *possible values* of X (e.g., if X is the result of a throw of a die, then its possible values are $1, 2, 3, 4, 5, 6$; if X is the number of radioactive decays before time t, so that $X = N(t)$, then its possible values are $0, 1, 2, 3, \ldots$). Either A is finite, and we can enumerate its elements, if there are n of them, as $a_1, a_2, \ldots, a_n$, or else

A is infinite and we can enumerate its elements as $a_1, a_2, \ldots$. Let us cover both cases by saying that A consists of elements a_i; $i = 1, 2, \ldots$, where the numbering is allowed to stop. We now define the *mean value* of X by the equation[19]

(i) $E(X) = \sum_i a_i \cdot P(X = a_i)$.

Prove that if B_k; $k = 1, 2, \ldots$ is a partition of Ω such that X is constant over each B_k (i.e., $\forall k \, \exists i \, \forall \omega \in B_k : X(\omega) = a_i$) then

(ii) $E(X) = \sum_k X(B_k) \cdot P(B_k)$.

Use this to prove that for two non-negative discrete random variables, defined on the same event space, the intuitively obvious formula

(iii) $E(X + Y) = E(X) + E(Y)$

holds.

Hint: Let a_i; $i = 1, 2, \ldots$ be the possible values of X. Furthermore, let $b_j; j = 1, 2, \ldots$ be the possible values of Y. Assume for the sake of simplicity that for all sample points ω, $X(\omega) \in \{a_i \mid i = 1, 2, \ldots\}$ and $Y(\omega) \in \{b_j \mid j = 1, 2, \ldots\}$ (or prove that one can allow oneself to make this assumption). Define a partition of Ω by looking, for each pair (i, j), at those sample points ω for which we have $X(\omega) = a_i$ and $Y(\omega) = b_j$.

Remark. If Ω is countable (a simplification the reader may allow himself to make), then things becomes easier, since we have the formula

(iv) $E(X) = \sum\limits_{\omega \in \Omega} X(\omega) \cdot P(\{\omega\})$

(cf. (ii)). It is easy to see that (iii) follows from (iv).

Exercise 10. Let $X \geq 0$ be a discrete random variable and let A be the set of possible values of X. Suppose that $A = \{a_1, a_2, a_3, \ldots\}$ where the a's are ordered so that $a_1 < a_2 < a_3 < \ldots$.

Draw a graph of the function

$$1 - F(x); \qquad x \geq 0.$$

where F is the distribution function of X.

Prove that

(i) $E(X) = \int_0^\infty (1 - F(x))dx$.

[19]When the sum here is not a convergent series, $E(X) = \infty$. Certainly, the considerations in this exercise also hold for an infinite mean value. But since such random variables do not appear in the concrete problems we have in mind, we ask the reader to assume that all random variables have finite mean values. For those who are interested, we mention that an example of a random variable with infinite mean value is the waiting time for a couple to have precisely the same number of male and female descendents. For another example, see Exercise 32

Exercise 11. Realize the reasonableness of the following formula for a non-negative random variable:

(i) $E(X) = \int_0^\infty (1 - F(x))dx = \int_0^\infty P(X > x)dx.$

Use the formula to compute the mean value of an exponentially distributed random variable and of a uniformly distributed random variable.

Remark: One can make more or less what one wants of this problem. E.g., one can aim to prove that if there is a mapping that associates to each non-negative random variable X a number $E(X)$ (where perhaps $E(X) = \infty$) such that the following axioms hold:

(ii) $E(X) =$ usual mean value for X discrete,
(iii) $X \leq Y \implies E(X) \leq E(Y)$,
(iv) $E(X + Y) = E(X) + E(Y)$,

then E is given by (i). To carry out this program one can define, for each value of n, the approximating discrete random variables Y_n and Z_n by

$$Y_n(\omega) = \frac{k}{n}$$

for all ω such that $\frac{k}{n} \leq X(\omega) < \frac{k+1}{n}$; $k = 0, 1, 2, \ldots$, and

$$Z_n(\omega) = Y_n(\omega) + \frac{1}{n}.$$

One difficulty which will be encountered in a more careful analysis is that, strictly speaking, it is not clear what the integrals in (i) should mean. If one has the energy, then one can clarify this problem by first *defining* the integrals in a suitable way, e.g., by an approximation process. The observant reader will already have noticed that (iv) is superfluous (a consequence of (ii) and (iii)).

Exercise 12. Prove that, for a non-negative random variable with finite mean value,

$$\lim_{x \to \infty} P(X > x) \cdot x = 0.$$

Hint: The area of the figure bounded by the x- and y-axes together with the function $y = P(X > x)$ is finite. Use geometric reasoning to conclude that there cannot be an $\varepsilon > 0$ such that

$$P(X > x) \cdot x \geq \varepsilon$$

for arbitrarily large x.

Next prove that, provided $X \geq 0$ has a density function f which is continuous on $[0, \infty[$, the following formula holds:

$$E(X) = \int_0^\infty x f(x) dx.$$

Hint. Use integration by parts on $\int_0^x P(X > t)dt$ and let $x \to \infty$. As a check, one can use the formula for an exponentially or uniformly distributed random variable.

Remark. The formula $E(X) = \int_{-\infty}^\infty x f(x) dx$ can also be used for a random variable that can assume negative values.

Exercise 13. For a random variable X with mean value μ, so that $E(X) = \mu$, define the *variance* $\sigma^2(X)$ as the mean value of the square of the difference between X and μ, i.e.,

$$\sigma^2(X) = E((X - \mu)^2).$$

The square root of the variance is called the *standard deviation* of X (and is denoted by $\sigma(X)$).

Prove that

$$\sigma^2(X) = E(X^2) - E(X)^2.$$

Prove that for a Poisson distributed random variable X with parameter λ,

$$E(X) = \sigma^2(X) = \lambda,$$

and for an exponentially distributed random variable with parameter λ,

$$E(X) = \sigma(X) = \frac{1}{\lambda}.$$

Exercise 14. Let X and Y be independent random variables. Realize that we have the formulae

(i) $E(X \cdot Y) = E(X) \cdot E(Y)$
(ii) $\sigma^2(X + Y) = \sigma^2(X) + \sigma^2(Y)$.

Illustrate, in a simple example, the importance of the assumption of independence.

Compute the variance of a binomially distributed random variable.

Hint. (ii) follows from (i), but (i) is not entirely easy to prove. Perhaps one can be content to notice that (i) always holds for success variables. It is more satisfactory to prove the formula for discrete random variables. It is convenient to introduce the notation 1_H for the success variable corresponding to the event H, i.e., $1_H(\omega) = 1$ for $\omega \in H$ and $1_H(\omega) = 0$ for $\omega \in H^c$. If X and Y are discrete, then they can be written in the form

$$X = \sum a_i 1_{A_i}, \qquad Y = \sum b_j 1_{B_j},$$

where the a_i's are the possible values of X and the b_j's are the possible values of Y. By multiplying these expressions one finds a formula for $X \cdot Y$ of the same type, involving certain success variables (in order to avoid "index confusion" one can first look at the case, say, where there are only two possible values for X and Y). By exploiting the independence, one is led to (i). The step from discrete to arbitrary random variables is in a certain sense not large (cf. Exercise 11).

Exercise 15. (i) Prove *Chebyshev's inequality*

$$P(|X - \mu| \geq \varepsilon) \leq \frac{\sigma^2(X)}{\varepsilon^2},$$

for any random variable with mean value μ and for any positive ε.

(ii) Prove the *law of large numbers* (in the so-called weak form): for any sequence $X_1, X_2, \ldots$ of independent, identically distributed random variables with mean value μ, and any positive ε, the probability

$$P\left(\left|\frac{X_1 + X_2 + \cdots + X_n}{n} - \mu\right| \geq \varepsilon\right)$$

converges to 0 as $n \to \infty$.

Investigate the content of this in the case where the X's are success variables.

Hint. (i): First prove that $E(Y) \geq \alpha \cdot P(A)$ for a non-negative random variable Y which satisfies the inequality $Y(\omega) \geq \alpha$ for all ω in the event A. Apply this with $Y = (X - \mu)^2$, $\alpha = \varepsilon^2$ and $A = \{| X - \mu | \geq \varepsilon\}$. (ii) follows easily from (i) and Exercise 14, (ii). [Strictly speaking, we have to assume $\sigma^2(X_1) < \infty$ for the proof just sketched to be valid.]

Exercise 16. Let v be an exponentially distributed random variable. Prove that, for all $s > 0$ and all $t > 0$,

(i) $$P(v \geq t + s \mid v \geq s) = P(v \geq t).$$

Prove that the latter property characterises the exponential distribution.

Hint. The function $t \mapsto -\log P(v \geq t)$; $t \geq 0$ is increasing and (i) gives a condition that links the function values at t, s and $t+s$ together.

Suppose that the waiting time for a certain event (e.g., a radioactive decay) is exponentially distributed. One makes 100 trials and finds that in about half of these one has to wait at least 10 seconds for the event to occur. Next one makes as many trials as are needed to have 100 cases where the waiting time is more than one minute. To get this, 6,000 to 7,000 trials must be made (think about it!). Show that one must expect that in about half of the 100 trials with waiting times over a minute, there will be a further 10 seconds' wait before the event occurs.

Remark. Property (i) and the above interpretation show that a system in which the waiting time is exponentially distributed behaves as though it were completely unaffected by the non-occurrence of an event. Thus there is no ageing effect or ripening effect. This property is precisely what we would expect of a spontaneous phenomenon. Thus the exercise is central to the theme of the book and can be said to express the fact that *the waiting time of a spontaneous event is exponentially distributed.*

Exercise 17. When we approximated observed data from a radioactive process by a Poisson distribution, we chose λ_{obs} to use as parameter (see §10). The object of this exercise is to show that the latter choice of parameter is the best possible, if we accept a certain overriding procedure, the *maximum-likelihood method.*

The starting point is a *sample* of *size* N from the Poisson distribution with unknown parameter λ. By this we mean that we are given N observed numbers

$$x_1, x_2, \ldots, x_N,$$

and that there are N random variables

$$X_1, X_2, \ldots, X_N,$$

which are independent and all Poisson distributed with the same parameter, and such that the x's are observed values of the X's, i.e., there is a sample point ω_0 such that[20]

$$x_1 = X_1(\omega_0), \quad x_2 = X_2(\omega_0), \quad \ldots, \quad x_n = X_N(\omega_0).$$

[20] A further elucidation of the sample concept is found in §17.

The problem is to give an *estimate* of the parameter λ or, as one says, an *estimator for* λ, on the basis of the observed numbers, and hence on the basis of the sample $x_1, \ldots, x_N$.

The idea is to compute, for *each* value of $\lambda > 0$, the quantities $L(\lambda)$ defined by

$$L(\lambda) = \text{probability that } X_1 = x_1, X_2 = x_2, \ldots, X_N = x_N$$
$$\text{given that the parameter is } \lambda.$$

In other words, $L(\lambda)$ gives the probability of a sample being precisely the one observed, when the true parameter is λ .

The function $\lambda \mapsto L(\lambda)$ is called the *likelihood function.* Thus the idea is that, once the concrete observed numbers $x_1, x_2, \ldots, x_N$ are given, the likelihood function depends only on λ.

If there is a unique value $\hat{\lambda}$ for which the likelihood function assumes its maximum value, then $\hat{\lambda}$ is called the *maximum likelihood estimator* for λ .

Write down a concrete expression for $L(\lambda)$.

Prove that the maximum likelihood estimator exists and is given by

$$\hat{\lambda} = \frac{x_1 + x_2 + \cdots + x_N}{N}.$$

Hint: Here, as in many other concrete situations, it is convenient to determine $\hat{\lambda}$ by looking for the minimum of the function $\lambda \mapsto -\log L(\lambda)$. In writing down an expression for this function, remember that each factor or addend which contains only expressions in the x's is considered to be a constant.

Prove that the estimator found is identical with the one given in formula (34).

Exercise 18. As is known, a sequence of numbers $a_1, a_2, \ldots$ is *convergent with limit* a if and only if for each $\varepsilon > 0$ there is an n_0 such that $|a_n - a| \le \varepsilon$ for all $n \ge n_0$. The condition says, roughly, that no matter how exactly one measures, from a certain point onwards the sequence can be reckoned to be constantly equal to a.

Prove that for any real x the sequence $(x^n/n!)_{n \ge 0}$ is convergent with limit 0:

$$\lim_{n \to \infty} \frac{x^n}{n!} = 0.$$

(The reader will be glad to have this result for Exercise 20.)

Exercise 19. The object of this exercise and the next one is to give an elementary understanding of infinite series and to prove a concrete result we used in the main text.

Let $a_0, a_1, \ldots$ be real numbers. The symbol $a_0 + a_1 + \cdots$ or briefly $\sum_{n=0}^{\infty} a_n$, or even more briefly $\sum_0^\infty a_n$, is called an *infinite series.* The a_n's are called the *terms* of the series. Associated with an infinite series $\sum_0^\infty a_n$ is a certain sequence of quantities called *partial sums,* denoted by $(s_n)_{n=0,1,2,\ldots}$ and defined by the formula

$$s_n = \sum_0^n a_k = a_0 + a_1 + \cdots + a_n; \qquad n \geq 0.$$

If the sequence (s_n) of partial sums is convergent with limit s, i.e. if $\lim_{n\to\infty} s_n = s$, then the infinite series $\sum_0^\infty a_n$ is said to be *convergent with sum s.* This is reflected by the notation

$$\sum_0^\infty a_n = s \qquad \text{or} \qquad a_0 + a_1 + a_2 + \cdots = s.$$

If $\sum_0^\infty a_n$ is convergent we shall accordingly regard $\sum_0^\infty a_n$ as a definite number, namely, the sum of the series.

Prove that the *geometric series* $\sum_0^\infty x^n$ is convergent for each value of x in the interval $-1 < x < 1$, and that the sum is given by

$$\text{(i)} \qquad 1 + x + x^2 + \cdots = \frac{1}{1-x}.$$

Remark. Perhaps some will be surprised at the effort involved in proving (i), and complain that (i) is actually trivial. This is because multiplying the series by x gives $x + x^2 + x^3 + \cdots$, and subtracting the latter from the series leaves only 1. The formula follows from this, and the reasoning makes no demands on x (except $x \neq 1$). E.g., we get $1 - 1 + 1 - 1 + \cdots = 1/2$ (setting $x = -1$). But one can see that the latter may not be so sound, since we also have

$$1 - 1 + 1 - 1 + 1 - 1 + \cdots = (1-1) + (1-1) + (1-1) + \cdots = 0$$

and yet again

$$\begin{aligned} &1 - 1 + 1 - 1 + 1 - 1 + 1 + \cdots \\ &\qquad = 1 + (-1+1) + (-1+1) + (-1+1) + \cdots = 1! \end{aligned}$$

And things go completely wrong when we try to apply the formula with $x = 2$.

Moral: One cannot make any deduction concerning operations involving infinitely many numbers solely on the basis of knowledge involving operations involving finitely many numbers. One must first specify the meaning of the operations involved and use the definitions thus set down to make a closer study of the properties of the new concepts.

Exercise 20. Prove that for each $n = 0, 1, 2, \ldots$ and for each real number x,

$$e^x = \sum_{k=0}^{n} \frac{x^k}{k!} + r_n(x),$$

i.e.,

$$e^x = 1 + \frac{x}{1!} + \frac{x^2}{2!} + \frac{x^3}{3!} + \cdots + \frac{x^n}{n!} + r_n(x),$$

where $r_n(x)$ is given by

$$r_n(x) = \frac{1}{n!} \int_0^x (x-t)^n e^t dt.$$

Then prove that for each real x,

$$e^x = \sum_0^\infty \frac{x^n}{n!}.$$

One says that $\sum_0^n x^k/k!$ is the nth *approximating polynomial* to e^x. Prove that the latter polynomial can be characterised as the polynomial of degree at most n that agrees with e^x for $x = 0$, and whose $1^{\text{st}}, 2^{\text{nd}}, \ldots, n^{\text{th}}$ derivatives at the point $x = 0$ also agree with those of e^x. The infinite series $\sum_0^\infty x^n/n!$ is called the *Taylor series* for e^x.

To compute e^x approximately one can use the nth approximating polynomial; the error committed in replacing e^x by the nth approximating polynomial is precisely $r_n(x)$, and one can give an estimate of its size. Do this, and use the result to compute e to two decimal places.

Those who would like to do so are invited to try to generalise the ideas in this exercise so that they are no longer confined to the exponential function. E.g., one can look into what happens with the functions $f(x) = \sin x$, $f(x) = \cos x$, $f(x) = \log(1+x)$ and $f(x) = (1-x)^\alpha$ (compare with Exercise 19 for $\alpha = -1$).

Exercise 21. Prove, without using Theorem 1, but directly from assumptions $A_1 - A_4$, that $E(N(I)) = s\lambda$ for each interval I of length s (formula (12)).

Hint: First assume that s is of the form n^{-1}, $n \in \mathbf{N}$, next that s is rational, and finally look at the general case.

Exercise 22. Prove that the regularity condition A_4 is satisfied for a Poisson process.

Hint: Verify equation (23).[21]

Exercise 23. In computing probabilities and mean values one is often pleased to have a method which may be called *computing by splitting up into possible causes*, also called the *law of total probabilities.* We take, as a starting point, a partition (B_j) of an event space Ω. The index j runs over a finite (or possibly countable) set. The event B_j is the "j^{th} cause." The B_j's are pairwise disjoint and fill up the whole of Ω.

For each event A and for each random variable X (with finite mean value) we have the formulae

$$P(A) = \sum_j P(B_j)P(A \mid B_j),$$

$$E(X) = \sum_j P(B_j)E(X \mid B_j).$$

Prove these, and show that the first formula can be regarded as a special case of the second.

Exercise 24. Let $(N(t))_{t>0}$ be a Poisson process with intensity λ and suppose that the events occurring in this process are detected with a detection probability p_{det}. Let $N_{\text{det}}(t)$ be the number of events detected in the time interval $]0, t]$.

Prove that $(N_{\text{det}}(t))_{t>0}$ is a Poisson process with intensity $p_{\text{det}} \cdot \lambda$.

---

[21] Of the other conditions, A_1 is the hardest to demonstrate. To illustrate the scope of A_1 we present the following. Let τ be a positive number. Define random variables K and V by $t_{K-1} \leq \tau < t_K$ and $V = t_K - \tau$. According to A_1, the random variable V is exponentially distributed with parameter λ and as such has mean value λ^{-1}. This seems paradoxical since the *whole* interval $[t_{K-1}, t_K]$ has length v_K and all the waiting times $v_1, v_2, \ldots$ surely have mean value λ^{-1}. For a detailed discussion of this and other "waiting time paradoxes" see Feller [10], vol. II, §I.4, where it is shown, among other things, that $E(v_K) = 2 \cdot \lambda^{-1}$!

Hint: Recall that (i), as well as (ii) and (iii) from Theorem 1 have to be verified. However, one may be content to derive (i). For this one can use Exercise 23, taking the "j^{th} cause" to be the event "$N(t) = k + j$"; $j = 0, 1, 2, \ldots$.

Exercise 25. Let $(N(t))_{t>0}$ be the superposition of the independent Poisson processes $(N_1(t))_{t>0}$ and $(N_2(t))_{t>0}$ with intensities λ_1 and λ_2 respectively.

Prove that $(N(t))_{t>0}$ is a Poisson process with intensity $\lambda_1 + \lambda_2$.

Exercise 26. Let $(N(t))_{t>0}$ be a Poisson process with intensity λ. Suppose $t < T$ and let K be a natural number. An obvious guess is that $N(t)$ is binomially distributed under the assumption $N(T) = K$; more precisely,

$$P(N(t) = k \mid N(T) = K) = \binom{K}{k} \cdot \left(\frac{t}{T}\right)^k \left(1 - \frac{t}{T}\right)^{K-k}; \quad k = 0, 1, \ldots, K.$$

Is this correct?

Remark. Indeed, this does prove to be correct. And one can easily arrange the proof so as to show the more general result that for an arbitrary interval $I \subseteq [0, T]$, of length t, $N(I)$ is binomially distributed under the assumption that $N(T) = K$. The parameters are $(K, t/T)$.

This situation can be obtained via another random model. Namely, let $X_1, X_2, \ldots, X_K$ be independent random variables which are all uniformly distributed over $[0, T]$. If we let $A(I)$, where $I \subseteq [0, T]$ is an interval of length t, denote the number of ν such that $X_\nu \in I$, then $A(I)$ is also binomially distributed with parameters $(K, t/T)$.

Comparing the two random models, an obvious conclusion is that, under the condition $N(T) = K$, the arrival times $t_1, t_2, \ldots, t_K$ of a Poisson process behave as independent uniformly distributed random variables over $[0, T]$, ordered by magnitude. One can give an exact proof for this conjecture. The result is important, since it gives an easy qualitative method for the statistical inspection of data where one considers a Poisson process to be the natural model – the step function $t \mapsto N(t)$ should run close to the straight line that connects $(0, 0)$ to $(T, N(T))$. If we wish for a quantitative test that exploits the result, then what we do is divide $[0, T]$ into suitable subintervals and carry out a χ^2-test; whether one starts from the position we have brought to light concerning uniform distribution, or whether one looks at the binomial distribution, is secondary. The expected as well as the observed numbers (N_k's and N_k^*'s in the notation of §17) are the same for both approaches.

Exercise 27. Let N be a fixed number and let $X_1, X_2, \ldots, X_N$ be random variables. We think of X_i as the number of particles in the "i^{th} region." The N regions we have in mind are assumed to be of the same size (e.g., they could be squares into which an object glass is divided, regions into which space is divided for making a star count, regions into which the sea is divided for counting animals or the like). Also let M be a random variable that represents the total number of particles in the N regions.

Suppose that M is Poisson distributed with parameter λ. Also suppose that the conditional distribution of the X's given M is a binomial distribution with parameters $(M, 1/N)$. This assumption can be formulated as follows: for each $m \in \{1, 2, \ldots\}$, for each $n \in \{0, 1, \ldots, m\}$ and for each $i \in \{1, 2, \ldots, N\}$,

$$P(X_i = n \mid M = m) = \binom{m}{n} \left(\frac{1}{N}\right)^n \left(1 - \frac{1}{N}\right)^{m-n}.$$

For $m = 0$, assume that $P(X_i = 0 \mid M = 0) = 1$.

Prove that the X_i's are Poisson distributed with parameter λ/N.

Hint: Use Exercise 23.

Remark. The above model does not determine all random properties of the X's. When we assume in addition that the X's are independent, then the model is completely determined.

Exercise 28. Let X be Poisson distributed with parameter λ. Determine k so that $P(X = k)$ is maximal.

Exercise 29. Let $(N(t))_{t>0}$ be a Poisson process with intensity λ. For real numbers $t_1 < t_2$ and non-negative whole numbers k_1, k_2 compute the probability

$$P(N(t_1) = k_1,\ N(t_2) = k_2).$$

Exercise 30. Let $(N(t))_{t>0}$ be a Poisson process. Find the mean value of the nth arrival time in terms of the intensity.

Exercise 31. Let $(N(t))_{t>0}$ be a Poisson process with intensity λ. For $\tau > 0$ and natural numbers K and N, derive a formula that makes it possible to compute the probability that at least K events occur in at least one of the N time intervals $[0, \tau)$, $[\tau, 2\tau)$, $\cdots$, $[(N-1)\tau, N\tau)$.

Then discuss how the probability in question can be regarded as the significance level for a test that investigates whether there has been too much "clustering" of events for the Poisson process model to be upheld.

Hint: Use the approach of §17, where the significance level for a test is the probability – computed under the assumption that the model is valid – of getting a sample at least as "extreme" as the sample actually observed.

In a concrete case 288 passages of motor cars on a motorway in the course of 37 minutes were observed (see Example 6). By dividing into 74 time intervals of length 1/2 minute it was found that the greatest number of car passages in any of these intervals was 14. Carry out a test for clustering as sketched above, and discuss whether one can uphold the Poisson process model, which of course rests mainly on an assumption of spontaneity.

Exercise 32. Let v_0, v_1, v_2, ... be independent non-negative random variables that all have the same continuous distribution. We think of the v's as "waiting times," but do not assume that they are exponentially distributed. More explicitly, we think of v_0 as "my" waiting time and of v_1, v_2, ... as my fellow citizens' waiting times; we imagine that citizens are visited at random and that the citizen first met has waiting time v_1, the next has waiting time v_2, etc.

Now let X be the random variable indicating the first citizen who is worse off than me, in the sense that his waiting time is greater than mine. More precisely, X is the least natural number n for which $v_n > v_0$.

Prove that the distribution of X is independent of the distribution of the v's, and that it satisfies the formula

$$\text{(i)} \qquad P(X = n) = \frac{1}{n(n+1)}; \qquad n = 1, 2, \ldots .$$

Hint: Express the event $X > n$ in terms of the random variables v_0, v_1, ... , v_n and compute $P(X > n)$.

Then prove that

$$\text{(ii)} \qquad E(X) = \infty.$$

Remarks: The result is paradoxical, in that it shows *I* am the world's unluckiest person, since on the average *I* have to look for an infinitely long time until I find a person worse off than myself. The reader can find this and other "waiting time paradoxes" discussed in more detail in Feller [10], vol. II. Feller emphasises that the irritating aspect of the above result is that one's fellow citizens can argue in exactly the same way that *they* are the world's unluckiest people! In connection with (i) it should be mentioned that the result can be used to make statistical tests of independence (since the distribution of the v's does not appear in (i)).

Exercise 33. Let $x_1, x_2, \ldots, x_n$ be a sample of size n from a distribution with mean μ and variance σ^2 (cf. Exercise 17). The sample *average* $\overline{x}$ and sample *variance* s^2 are defined by

$$\overline{x} = \frac{1}{n}\sum_{i=1}^{n} x_i, \qquad s^2 = \frac{1}{n-1}\sum_{i=1}^{n}(x_i - \overline{x})^2.$$

Using the obvious notation, the corresponding random variables are given by

$$\overline{X} = \frac{1}{n}\sum_{i=1}^{n} X_i, \qquad S^2 = \frac{1}{n-1}\sum_{i=1}^{n}(X_i - \overline{X})^2.$$

(i) Prove that

$$s^2 = \frac{1}{n-1}\sum_{i=1}^{n} x_i^2 - \frac{n}{n-1}\,\overline{x}^2.$$

(ii) Prove that $\overline{x}$ and s^2 are *central estimators* of μ and σ^2 respectively, i.e.,that

$$E(\overline{X}) = \mu, \qquad E(S^2) = \sigma^2.$$

(iii) The *Lexis coefficient* q^2 of the sample $x_1, x_2, \ldots, x_n$ is defined by

$$q^2 = s^2/\overline{x}.$$

The corresponding random variable is denoted by Q^2.[22]

[22] Wilhelm Lexis, a German economist, introduced this coefficient in 1877 (instead of q^2 he used $(n-1)n^{-1}q^2$). If the sample is from a Poisson distribution, then one can show that, provided the parameter λ is not too small, $(n-1)Q^2$ is approximately χ^2-distributed with $n-1$ degrees of freedom. If n is large one can show that $(Q^2 - 1)\sqrt{\frac{1}{2}(n-1)}$ is approximately normally distributed. These circumstances give rise to a test for Poisson distribution (compare with §17). If the sample is from a modified Poisson distribution Bortkiewicz used the formula (notation as in §16)

$$\hat{h} = TUV^{-1}(1 + 1/\gamma)^{-1}, \text{ where } \gamma = UV^{-1}\left(q^2 - \sqrt{q^4 - VU^{-1}(1-q^2)}\right)$$

as an estimator for the dead time (the formula is only usable when $q^2 \approx 1$).

Show that, provided the sample is from a Poisson distribution, one must expect that $q^2 \approx 1$, and if the sample is from a modified Poisson distribution, one must expect that $q^2 < 1$.

(iv) We say that a random variable X is a (finite) *mixture of Poisson distributions* if there is a random variable Z, which can assume only finitely many values, such that for each z with $P(Z = z) > 0$ the conditional distribution of X given that $Z = z$ is a Poisson distribution, i.e., there is a λ corresponding to z such that

$$P(X = k \mid Z = z) = \frac{\lambda^k}{k!}e^{-\lambda}; \qquad k = 0, 1, 2, \dots .$$

Show that, provided that the sample $x_1, x_2, \dots, x_n$ is from such a mixed distribution, one must expect that $q^2 > 1$.

Hint. Find a formula for $\sigma^2(X)/E(X)$ by using Exercise 23 to compute $E(X)$ and $E(X^2)$. The inequality $\sigma^2(X)/E(X) \geq 1$ can be derived by a geometric argument concerning the graph of the function $\lambda \mapsto \lambda^2$. (Perhaps limit the investigation to the case where Z can assume only two values.) For appropriate numerical material we refer to Example 12.

(v) We now assume that all numbers in the sample are natural numbers or 0 and we introduce the notation from §10 and from program P1:

$$N_k = \text{ number of } i \text{ such that } x_i = k; \qquad k = 0, 1, 2, \dots ,$$
$$U = \sum N_k, \qquad V = \sum kN_k, \qquad W = \sum k^2 N_k.$$

Express q^2 in terms of U, V and W and suggest an extension of P1 that also computes q^2 (those interested can exploit the information in the footnote to make a more thorough extension of P1).

Exercise 34. Let $X_1, X_2, \dots, X_N$ be independent random variables, all Poisson distributed with unknown parameter λ. Let $\hat{\lambda}$ denote the maximum likelihood estimator of λ, i.e. (see Exercise 17),

$$\hat{\lambda} = \frac{X_1 + X_2 + \cdots + X_N}{N}.$$

Determine $\sigma(\hat{\lambda})$ and show that the demand "$\sigma(\hat{\lambda}) \leq a\%$ of λ" is equivalent to $\lambda \cdot N \geq 10^4/a^2$.

Hint: Use Exercise 14(ii).

Exercise 35. Let $(N(t))_{t>0}$ be a Poisson process with intensity λ, which serves as a model of radioactive decay. The phenomenon is studied with the help of a counter with dead time h. Model I from §12 is used, i.e., only registered decays cause paralysis. We write

$$N(t) = N_{\text{reg}}(t) + N_{\text{unreg}}(t),$$

where the first term is the number of registered decays in $]0,1]$ and the second is the corresponding number of unregistered decays.

The *registered intensity* is defined by

$$\lambda_{\text{reg}} = E(N_{\text{reg}}(t))/t.$$

(We take it for granted that only time points t much larger than h are considered, so that λ_{reg} can be reckoned to be constant. Then there is also agreement with (39), §16.)

(i) Assuming it is known that $N_{\text{reg}}(t) = n$, realize that the mean number of unregistered decays in $]0,t]$ is given to very good approximation by

$$E(N_{\text{unreg}}(t) \mid N_{\text{reg}}(t) = n) = nh\lambda$$

and use this to establish that

$$E(N(t)) = E(N_{\text{reg}}(t)) \cdot (1 + h\lambda).$$

From the latter, derive the equations

$$\lambda = \lambda_{\text{reg}}(1 - h\lambda_{\text{reg}})^{-1} = (1/\lambda_{\text{reg}} - h)^{-1}. \tag{51}$$

Remark. The formula (51) follows in more primitive fashion from the following argument: There are approximately λ_{reg} registered decays per unit time. These cause paralysis of the counter for $\lambda_{\text{reg}}h$ units of time, and in the latter amount of time we expect $\lambda\lambda_{\text{reg}}h$ decays. Therefore $\lambda_{\text{reg}} + \lambda\lambda_{\text{reg}}h = \lambda$. However, the derivation suggested above permits a more detailed analysis if one so desires.

(ii) We now suppose that a certain number – let us say k – of Poisson processes are observed over time intervals of length t. Call the intensities λ^i; $i = 1, 2, \ldots, k$ and the registered intensities λ^i_{reg}; $i = 1, 2, \ldots, k$. The superposition Poisson process is also observed over a time interval of length t; the intensity and registered intensity for the superposition are called λ^s and λ^s_{reg} respectively. Prove that

$$(1/\lambda^s_{\text{reg}} - h)^{-1} = \sum_{i=1}^{k} (1/\ \lambda^i_{\text{reg}} - h)^{-1}.$$

(iii) (*Dead time determination by the two source method.*) Let $k = 2$ and consider observations originating from counting with 1) the radioactive source A, 2) the two radioactive sources A and B together and 3) the radioactive source B. [For purely practical reasons it is best to arrange the series of observations as indicated.]

Derive the formula

$$h = \frac{1}{\lambda^{12}_{\text{reg}}} \left(1 - \sqrt{1 - \lambda^{12}_{\text{reg}} \frac{\lambda^1_{\text{reg}} + \lambda^2_{\text{reg}} - \lambda^{12}_{\text{reg}}}{\lambda^1_{\text{reg}} \lambda^2_{\text{reg}}}} \right), \tag{52}$$

where we have set $\lambda^{12}_{\text{reg}} = \lambda^s_{\text{reg}}$. Since the λ_{reg}'s can be estimated as the averages of observed values of $N_{\text{reg}}(t)/t$, (52) can be used to give an estimator of the dead time h. Think about it!

Remarks. One often sees the approximate formula

$$h = \frac{\lambda^1_{\text{reg}} + \lambda^2_{\text{reg}} - \lambda^{12}_{\text{reg}}}{2\lambda^1_{\text{reg}} \lambda^2_{\text{reg}}} \tag{53}$$

quoted instead of (52). Since (52) is no more troublesome than (53) – in the days of the pocket calculator – and since (53) is subject to great inaccuracy for realistic data sets (see, e.g., the numbers in Table 6!), one should always use (52).

Whichever formula one uses, one must remember to correct (the estimates for) the registered intensities by subtracting (an estimate for) the background intensity.

One can refine these considerations by noting that in reality the observations are: 1) A + background, 2) $A + B$ + background, and 3) B + background. Using an added superscript 0 to indicate that the background is also taken into consideration, one can derive the formula

$$h = \frac{p_1 - p_2}{p_1 s_2 - p_2 s_1} \left(1 - \sqrt{1 - \frac{(p_1 s_2 - p_2 s_1)(s_1 - s_2)}{(p_1 - p_2)^2}} \right), \tag{54}$$

where

$$p_1 = \lambda^{10}_{\text{reg}} \cdot \lambda^{20}_{\text{reg}}, \quad p_2 = \lambda^{120}_{\text{reg}} \cdot \lambda^{0}_{\text{reg}},$$
$$s_1 = \lambda^{10}_{\text{reg}} + \lambda^{20}_{\text{reg}}, \quad s_2 = \lambda^{120}_{\text{reg}} + \lambda^{0}_{\text{reg}}.$$

Those who are keen are invited to derive (54). Although the latter formula is more precise than (52), their difference for "normal" data sets is not of practical significance.

Exercise 36. The object of this exercise is to give a way of setting up a two-source experiment to determine the dead time so that one can expect a certain precision. For a description of the method, see §12 and also Exercise 35.

Let X_1, X_2 and Y indicate the numbers of registered decays for source 1, source 2, and source 1 and source 2 together, respectively. Suppose that the observations are over τ seconds. Let λ_1, λ_2 and μ denote the corresponding registered intensities, i.e., $\lambda_1 = X_1/\tau$, $\lambda_2 = X_2/\tau$ and $\mu = Y/\tau$. Assume that λ_1 and λ_2 have approximately the same size λ.

Let $\hat{h}$ denote the estimator of the dead time h according to the two-source method. Show that the demand "$\sigma(\hat{h}) \leq b\%$ of h" is roughly equivalent to

$$\tau \geq h^{-2}\lambda^{-3}b^{-2}10^4. \tag{55}$$

Hint:

$$\hat{h} \approx \frac{\lambda_1 + \lambda_2 - \mu}{2\lambda_1\lambda_2} \approx \frac{1}{2\tau\lambda^2}(X_1 + X_2 - Y) \approx \frac{1}{2\tau\lambda^2}(X_1 + X_2 - Y_1 - Y_2),$$

where X_1, X_2, Y_1 and Y_2 are independent, Y_1 is distributed approximately as X_1, and Y_2 is distributed approximately as X_2.

Remarks: The inequality (55) is a guide. The value of h that should be used on the right-hand side only needs to be a rough estimate. One should not try to satisfy (55) by simply choosing λ large (bringing the sources close to the counter), since our dead time model is inaccurate in this case. One should have λh distinctly less than 1 (e.g., $\lambda h < 0.05$).

Exercise 37. As in the previous exercise, we are concerned with a radioactive preparation that gives rise to decays that are registered on a counter with dead time h. We use Model I from §12.

(i) Show that

$$t_n = \tau_n + (n-1)h; \qquad n = 1, 2, \ldots$$

where the notations are those introduced in Figure 21, and also explain why $\tau_1, \tau_2, \ldots$ are the arrival times for a Poisson process with intensity λ (the same intensity which represents all events, both registered and unregistered (see Figure 21, part a)).

(ii) As in the previous exercise, let $N_{\text{reg}}(t)$ denote the number of registered events in $]0,t]$. Use (i) together with Theorem 2 to prove that, for $nh < t$,

$$P(N_{\text{reg}}(t) \leq n) = e^{-\lambda(t-nh)} \sum_{k=0}^{n} \frac{(\lambda(t-nh))^k}{k!}. \tag{56}$$

Remark. Naturally, (56) can be used to set up a formula for the quantity $P(N_{\text{reg}}(t) = n)$. This becomes somewhat complicated, among other reasons, because one must split into 3 cases: $nh < t$, $(n-1)h < t < nh$, and $t \leq (n-1)h$. In applications one will normally have nh less than t, and one can then arrange to use only (56).

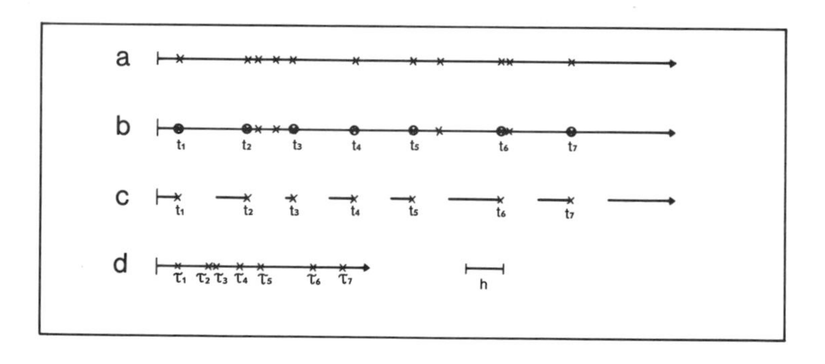

Fig. 21.

a: All events; described by a Poisson process with intensity λ.
b: Marking of registered events and corresponding arrival times $t_1, t_2, \ldots$.
c: Registered events with empty spaces for "dead" periods.
d: Compression of registered events by neglecting dead periods; compressed arrival times: $\tau_1, \tau_2, \ldots$.

Exercise 38. In this exercise we set up a model, which is in fact more accurate than the one in the main text, for radioactive decay. It allows us to model intensity that varies with time, making it possible to discuss the half-life of the preparation in question.

We consider a radioactive preparation which at time 0 consists of $2N$ atoms which we imagine to be numbered; in practice one must imagine N to be of the order of Avogadro's number, around 10^{23}. Associated with the preparation we have $2N$ random variables $V_1, V_2, \ldots, V_{2N}$. The random variable V_k gives the time of decay of the kth atom, and hence the waiting time until the kth atom transforms. We call the V's *individual waiting times.*

It is assumed that $V_1, V_2, \ldots, V_{2N}$ are independent, identically distributed random variables, each with an exponential distribution. The parameter is denoted by λ_{ind} and is called the *decay constant*; its dimension is time^{-1}. It should be mentioned that we assume each one of the $2N$ atoms to contribute only once to the radioactive process, namely, at its time of decay, and hence at time V_k for the kth atom. One can think, e.g., of one of the radioactive processes involving a proper nuclear transformation (typically an α- or β-active preparation) where the new nucleus is stable.

The assumption that the V's are independent and identically distributed can certainly be accepted immediately. It should also seem reasonable that the distribution is exponential, from the arguments in the main text, and this can also be supported by the result of Exercise 16.

(i) Define the *half-life* $T_{1/2}$ by the equation

$$P(V_1 \leq T_{1/2}) = P(V_1 \geq T_{1/2})$$

and show that

$$T_{1/2} = \frac{\log 2}{\lambda_{\text{ind}}}.$$

(ii) Define the arrival times $t_1, t_2, \cdots, t_{2N}$ as

$$\begin{aligned} t_1 &= \text{ smallest of } V_1, V_2, \ldots, V_{2N}, \\ t_2 &= \text{ second smallest of } V_1, V_2, \ldots, V_{2N}, \\ &\vdots \\ t_{2N-1} &= \text{ second largest of } V_1, V_2, \ldots, V_{2N}, \\ t_{2N} &= \text{ largest of } V_1, V_2, \ldots, V_{2N}. \end{aligned}$$

Express the event $\{t_1 \geq s\}$, where s is a fixed positive number, in terms of the random variables $V_1, V_2, \ldots, V_{2N}$. Prove that t_1 is exponentially distributed with parameter $2N\lambda_{\text{ind}}$.

(iii) Define waiting times (the *global waiting times* or *population waiting times*) as usual, i.e.,

$$v_1 = t_1, \quad v_2 = t_2 - t_1, \quad \cdots, \quad v_{2N} = t_{2N} - t_{2N-1}.$$

Use the result of (ii) to show that the waiting times are independent and exponentially distributed, and show that the parameter for v_k's distribution is $(2N - k + 1)\lambda_{\text{ind}}$.

(iv) The *stochastic half-life* is defined to be t_N. Prove that the mean of the latter is

$$E(t_N) = \frac{1}{\lambda_{\text{ind}}}\left(\frac{1}{N+1} + \frac{1}{N+2} + \cdots + \frac{1}{2N}\right).$$

Compare with the result under (i) (remember that N is very large!)

(v) How much longer should we expect to wait, on average, for all atoms to be transformed, compared with the average time for half the atoms to be transformed, e.g., for $N = 10^{23}$? (Compare, say, with $E(t_{2N}/t_N)$, although the latter is more difficult to compute exactly.)

(vi) Prove that the variance of t_N is

$$\sigma^2(t_N) = \frac{1}{\lambda_{\text{ind}}^2}\left[\frac{1}{(N+1)^2} + \frac{1}{(N+2)^2} + \cdots + \frac{1}{(2N)^2}\right]$$
$$\approx \frac{1}{\lambda_{\text{ind}}^2} \cdot \frac{1}{2N}$$

and give an estimate of the ratio between the standard deviation and the mean of t_N. Does the ratio found conform with what one normally teaches – that the part of a radioactive preparation still active is a deterministic quantity, which declines exponentially with time? Compare, say, with Thesis 10.

Exercise 39. We consider a system (a "radioactive soup") in which a series of radioactive processes take place. At each moment the system consists of a certain number of atoms, some of "type 1," some of "type 2," etc. The number of atoms of a particular type k in the system at time t is denoted by $x_k(t)$.

The atoms of type k are radioactive; the corresponding decay constant is denoted by λ_k and the half-life by T_k (cf. Exercise 38). No particular assumptions about the kind of decay products are made in this exercise. Normally the decays will be accompanied by radiation. It should be pointed out that we allow some of the atoms in the system to be stable. When the atoms of type k are stable, this merely corresponds to $\lambda_k = 0$ and $T_k = \infty$. The other extreme, where $T_k \approx 0$, can also occur. This corresponds to a very unstable atom, which decays almost immediately.

The *activity* at time t of the atoms of type k is denoted by $A_k(t)$ and defined by

$$A_k(t) = x_k(t) \cdot \lambda_k.$$

Show the following: the decays of type k atoms in the system in a time interval $[t, t+\tau]$, where τ is sufficiently small, constitute a Poisson process, whose intensity is precisely the activity $A_k(t)$ at time t. In comparison to what should τ be small, if this is to hold?

Make suitable assumptions about the accompanying radiation and explain how one can in principle determine the activities $A_1(t), A_2(t), \ldots$ on the basis of observations of radiation from the system.

Exercise 40. We now assume that the system of Exercise 39 is a *cascade process*:

$$x_1 \xrightarrow{\lambda_1} x_2 \xrightarrow{\lambda_2} x_3 \xrightarrow{\lambda_3} x_4 \xrightarrow{\lambda_4} \cdots. \tag{57}$$

This means that for each $k = 1, 2, \ldots$ the atoms of type k have only atoms of type $k+1$ as decay product (perhaps together with accompanying radiation, which does not play a role in this exercise), and the decay of one type k atom gives precisely one type $k+1$ atom.

The cascade process "breaks off" if there is a k with $\lambda_k = 0$.

The viewpoint we shall take in this exercise is purely deterministic. It is evident from Exercise 38 that this viewpoint is compatible with the fact that the process is actually stochastic. From the latter exercise we have the following result:

$$x_1(t) = x_1(0)e^{-\lambda_1 t}; \qquad t \geq 0$$

(the latter formula can be derived by purely deterministic considerations by finding the derivative of x_1; such a derivation naturally does not give nearly as much insight as the approach in Exercise 38).

(i) For the cascade process $x \xrightarrow{\lambda} y \xrightarrow{\mu} z$, suppose that $y(0) = 0$ and that the function $t \mapsto x(t)$; $t \geq 0$ is known. Prove that the function $t \mapsto y(t)$; $t \geq 0$ can then be determined by

$$y(t) = \lambda e^{-\mu t} \int_0^t x(\tau) e^{\mu\tau} d\tau; \qquad t \geq 0.$$

Hint: Use that $y(t) = \int_0^t$ {the amount of y produced in the time interval $[\tau, \tau + d\tau]$ which is still "alive" $t - \tau$ time units later, i.e., at time t}$d\tau$.

(ii) Assuming $x(t) = \alpha e^{-\beta t}$; $t \geq 0$ where α and β are positive constants, show that

$$y(t) = \alpha \cdot \frac{\lambda}{\mu - \beta}(e^{-\beta t} - e^{-\mu t}); \qquad t \geq 0,$$

for the cascade process in (i).

(iii) Suppose that $x_2(0) = x_3(0) = \cdots = 0$ for the cascade process (57). Prove that for each $k = 1, 2, \ldots$ the function $t \mapsto x_k(t)$; $t \geq 0$ is a sum of functions of the type $t \mapsto \alpha e^{-\beta t}$; $t \geq 0$. For $k = 1, 2, 3$ one should also prove the following formulae, all valid for $t \geq 0$:

$$x_1(t) = x_1(0)e^{-\lambda_1 t}, \tag{58}$$

$$x_2(t) = x_1(0)\frac{\lambda_1}{\lambda_2 - \lambda_1}\left[e^{-\lambda_1 t} - e^{-\lambda_2 t}\right], \tag{59}$$

$$\begin{aligned} x_3(t) = x_1(0)\lambda_1\lambda_2 \Bigg[& \frac{1}{(\lambda_2 - \lambda_1)(\lambda_3 - \lambda_1)} e^{-\lambda_1 t} \\ & - \frac{1}{(\lambda_2 - \lambda_1)(\lambda_3 - \lambda_2)} e^{-\lambda_2 t} + \frac{1}{(\lambda_3 - \lambda_1)(\lambda_3 - \lambda_2)} e^{-\lambda_3 t} \Bigg]. \end{aligned} \tag{60}$$

(iv) For a cascade process (57), determine the type of the function $t \mapsto x_1(t) + x_2(t) + \ldots$.

Remark. It is recommended to control the formulas (58)–(60) by investigating the cases $\lambda_1 = 0$, $\lambda_2 = 0$ and $\lambda_3 = 0$.

Exercise 41. Consider the cascade process (57) with

$$x_2(0) = x_3(0) = \cdots = 0.$$

Prove that the activity of type 2 atoms is maximal at time $t_{\max}$ given by

$$t_{\max} = \frac{\log \lambda_2 - \log \lambda_1}{\lambda_2 - \lambda_1} \tag{61}$$

and compare the activities $A_1(t) = \lambda_1 x_1(t)$ and $A_2(t) = \lambda_2 x_2(t)$ at that timepoint.

In practice one will never have $\lambda_1 = \lambda_2$, but even when $\lambda_1 \approx \lambda_2$ formula (61) is inconvenient. Find a reasonable formula that covers the latter case.

Exercise 42. Consider again the cascade process (57) with

$$x_2(0) = x_3(0) = \cdots = 0.$$

Suppose that $T_1 > T_2$ (equivalently, $\lambda_1 < \lambda_2$).

Investigate, for large values of t, the ratio $A_2(t)/A_1(t)$ between the activities of the "daughter" and "mother" atoms.

Give diagrams of the functions $t \mapsto A_1(t)$ and $t \mapsto A_2(t)$ in the case where T_1 is merely a bit larger than T_2 , and in the case where T_1 is much larger than T_2 (e.g., in the latter case take $T_1 = 30$ years and $T_2 = 2.6$ min., corresponding to the process in Example 14).

Exercise 43. For the cascade process (57), assume that $x_2(0) = 0$. Set

$$\kappa(t) = \frac{x_2(t)}{x_1(t)}.$$

Show that the function $t \mapsto \kappa(t)$; $t \geq 0$ has an inverse. Determine the latter, and discuss how this result can be used for age determinations (the method most commonly used is in the case $\lambda_2 = 0$).

Exercise 44. For the radioactive process $x_1 \xrightarrow{\lambda} x_2$, where the type 2 atoms are stable, show that

$$x_2(t) = x_2(0) + (e^{\lambda t} - 1)x_1(t).$$

For a certain specimen (e.g., a piece of rock) the quantities $x_1(t)$ and $x_2(t)$ are measured at time t. Show how one can determine t when one also knows $x_2(0)$. (Example 15 discusses a possible way of proceeding when $x_2(0)$ is not known.)

Exercise 45. The excerpt below is from *Samvirke*, August 1980. It is part of an article by professor Ove Nathan on modern atomic physics.

The age of the universe – a brief second.

Recently the quark physics have questioned yet another "sacred cow": the absolute stability of the proton. The quark model predicts that the proton is radioactive with a half-life around 10^{30} *years (a one followed by 30 zeros). The half-life is the time which must elapse before half of a given number of protons spontaneously decay to other particles. Of course one cannot directly measure a time interval of* 10^{30} *years – the age of the universe itself (around 20 billion years) is only a brief moment in comparison. But if one has many protons, then in the course of one year one can hope to observe the decomposition of a few of them. The experiment is within the realm of possibility if one observes around* 10^{30} *protons, which corresponds to the number of protons in a medium-sized swimming pool full of water.*

In the coming months the first experimental attempt to measure the spontaneous decomposition of the proton will be set up in an abandoned mineshaft deep in the Earth.

We assume from now on the validity of the quark model, according to which the proton is radioactive with a half-life of 10^{30} years.

Give an expression for the function f which describes how many protons remain at time t, measured in years, when there are 10^{30} protons at time 0.

Find a first degree approximating polynomial to f at the point 0.

Use this to make a judgement about how many of the 10^{30} protons decompose in the first year, and compare with the information in the article.

The article presupposes a deterministic model. Is this reasonable? Discuss the conclusions of the article from a stochastic viewpoint.

Exercise 46. When radiation passes through matter an *attenuation* occurs, i.e., a weakening of energy. The easiest to study is γ-radiation, for which a γ-quantum passing through matter at each moment either remains unaltered or else undergoes an interaction in which the γ-quantum vanishes.

Making suitable assumptions, show that there is a constant μ, the *attenuation constant*, such that the intensity I is expressed as a function of the distance d traversed through the material as

$$I(d) = I(0)e^{-\mu d}.$$

Define the *half-distance* d_* by $I(d_*) = \frac{1}{2}I(0)$ and find a relation between μ and d_*.

Examples for Further Investigation

The examples below normally do not end with the formulation of a series of questions. In each case the problem is to investigate whether a stochastic model can be set up to "explain" observations. Often there are several ways to check a model, cf. §§16, 17.

It is left to the reader to decide in each individual case how much basis there is for the reasonableness of a stochastic model, and how much can be made of the statistical analysis. Many of the exercises contain information that can be used.

We remark that for some of the examples most readers will probably come to the conclusion that the most obvious model must be rejected.

Over and above the mathematical treatment of the examples it is important that one try to familiarise oneself with the phenomenona that are the subject of the modelling being carried out. For some of the examples a closer examination of source material would be in order. E.g., for Example 3 one could learn why collection was made from different depths, whether there are creatures other than water fleas, etc.

Example 1. (see [10], vol.I). During World War II, some inhabitants of London were surprised that their neighbourhood was hit more frequently by German flying bombs compared with other neighbourhoods, despite the fact that the bombs were fired from a long way off and, one would have thought, the Germans fired the bombs merely on a course for Greater London.

The whole of South London was divided into 576 small regions, each $\frac{1}{4}$ square kilometer, and the number of hits in each region was counted. Table 10 shows how many regions received 0, 1, 2, 3, 4 or 5 or more hits:

Table 10 Bombs over London

0	1	2	3	4	≥ 5
229	211	93	35	7	1

Example 2. In auto insurance records for a certain Danish company the number of accidents for the year 1978 was worked out. For each insured it was noted how many accidents the person in question was involved in, and these accidents were divided into self-inflicted accidents and accidents caused by others.

Table 11 Car accidents

S \ O	0	1	2
0	566	44	2
1	55	4	3
2	3	1	1
3	1	2	

The results are shown in Table 11. We see from the table that 3 policy holders had one self-inflicted and 2 non-self-inflicted accidents, and likewise that there were 3 policy holders with 2 self-inflicted and 0 non-self-inflicted accidents, in the year in question.

In a more detailed investigation of these data it would be natural to investigate self-inflicted and non-self-inflicted accidents separately. One should also set up a model for the total data on the assumption that the two types of accident are independent.

Example 3. From the ice-floe station "Fram 1", biologist Lars Haumann collected water fleas from the polar ocean between Greenland and Svalbard in April 1979. Two different kinds were found, *Calanus hyperboreus* and *Calanus glacialis* (a few examples of *Calanus finmarchicus*, which are difficult to distinguish morphologically from *Calanus glacialis*, may appear). Both kinds can occur at four stages of development, denoted by A, B, ♀ and ♂.

At each of the depths 0-35 m, 35-125 m, 125-250 m and 250-300 m, 12 samples were taken. The results are given in Table 12, which shows the number of individuals in each category in each sample (each of the 12 columns corresponds to 4 subsamples). It should perhaps be added that the samples in columns 1-3 were taken during the time 12.30 - 17.30, the samples in columns 4-6 during 18.45 - 23.45, the samples in columns 7-9 during 00.30 - 05.00, and the samples in columns 10-12 were taken during 06.15 - 11.45.

Table 12 Collection of water fleas from the polar sea

			1	2	3	4	5	6	7	8	9	10	11	12
0.–35 m	*hyperboreus*	A							1	1		3		
		B										1		
		♀												
		♂												
	glacialis	A												
		B												
		♀		2	2	2	1		1	1				
		♂				1								
35–125 m	*hyperboreus*	A		1		3	1			3	2		2	1
		B	3	3	4	2	8	5	3	5	4		3	3
		♀	6	9	7	6	7	3	8	7	5	3	6	8
		♂												
	glacialis	A	1	1	3	3	1		6			1	2	
		B	13	13	22	15	16	11	17	9	19	6	14	9
		♀	65	80	75	66	44	39	72	49	52	46	48	49
		♂	1	1	5	1	1		2	1	3	2	1	3
125–250 m	*hyperboreus*	A	26	14	15	24	20	25	24	24	18	21	35	20
		B	16	19	12	20	16	20	19	20	25	11	22	21
		♀	25	29	10	21	15	31	23	22	24	20	29	23
		♂		2					1					1
	glacialis	A				1			2					
		B		2		3	1	2	3	1	2	5	4	3
		♀	1	5	5		2	4	2	2	1	3	6	
		♂		1					1					1
250–300 m	*hyperboreus*	A	12	15	10	7	3	7	3	3	6	3	1	2
		B	3	5	4	6	8	5	4	3	8	3	9	4
		♀	6	7	3	1	6	4		5	7	5	4	4
		♂			1					1	1			2
	glacialis	A												
		B		1								2		
		♀	1	2	3		1				2			
		♂	1		1									

Example 4. (from [1]). A solution of bacteria is given, containing an unknown number of mutants that ferment maltose. Among 100 samples of 0.1 ml, one finds 58 that do not ferment, and on the basis of this one wishes to estimate the average number of mutants per ml.

Remark: One cannot distinguish completely between numbers of mutants, only between the two categories "0 mutants" and "more than 0 mutants."

Example 5. (from [1]). A type of *coli*-bacterium is assumed to have a certain probability of mutating to a streptomycin-resistant form.

150 petri dishes with streptomycin agar are each supplied with 1 million coli bacteria. If a bacterium mutates, it either produces a small colony which survives, or else it is killed by the streptomycin.

The numbers of dishes with 0, 1, 2, 3 and 4 colonies of bacteria respectively are given in Table 13.

Table 13

0	1	2	3	4
98	40	8	3	1

The discussion of this example should also contain an estimate of the mutation probability.

Example 6. Table 14 shows the result of a traffic count of south-bound cars on the motorway from Elsinore to Copenhagen. The count was carried out at Kokkedal, about 14.5 km south of Elsinore, on Friday, February 18, 1983 between about 10.30 and 11. The table shows the times at which cars passed; thus we see that the last passage (no. 288) occurred 36 minutes and 58 seconds after the beginning of observations. The table also shows the waiting times.

It is in fact easy to set up a traffic count – if the road and time are chosen suitably – so that one can expect to find a high degree of spontaneity in the data (Poisson process!). The example stems from a count carried out by students of Rungsted State School, and one of the objectives of the count was the processing of observations undertaken by oneself, in place of more official numerical material which was given in the Danish edition of the present book.

It should be mentioned that many students noted that a series of cars which passed in the 29th minute after the start of observations were

Table 14 Time for passages of cars

Time		Time		Time		Time		Time		Time	
00 09	9	07 59	32	14 19	11	19 42	0	26 07	12	31 29	3
00 18	9	08 05	6	14 33	14	19 44	2	26 15	8	31 31	2
00 28	10	08 11	6	14 36	3	19 56	12	26 15	0	31 37	6
00 33	5	08 14	3	14 40	4	19 58	2	26 19	4	31 52	15
00 35	2	08 24	10	14 58	18	20 03	5	26 20	1	31 59	7
00 42	7	08 25	1	15 00	2	20 20	17	26 22	2	32 04	5
00 43	1	08 28	3	15 05	5	20 31	11	26 53	31	32 12	8
00 50	7	08 44	16	15 25	20	20 33	2	27 12	19	32 15	3
00 51	1	08 55	11	15 51	26	20 38	5	27 17	5	32 32	17
01 02	11	08 56	1	15 59	8	20 41	3	27 20	3	32 40	8
01 13	11	08 58	2	16 4	5	20 47	6	27 22	2	32 42	2
01 18	5	08 58	0	16 12	8	20 49	2	27 39	17	32 48	6
01 46	28	09 01	3	16 13	1	20 50	1	28 03	24	33 04	16
02 03	17	09 06	5	16 17	4	21 05	15	28 14	11	33 10	6
02 13	10	09 12	6	16 20	3	21 33	28	28 18	4	33 15	5
02 25	12	09 15	3	16 33	13	21 35	2	28 21	3	33 15	0
02 27	2	09 17	2	16 41	8	21 36	1	28 26	5	33 25	10
02 33	6	09 45	28	16 41	0	21 37	1	28 26	0	33 34	9
02 38	5	10 07	22	16 45	4	21 44	7	28 43	17	33 48	14
03 03	25	10 11	4	16 48	3	21 53	9	28 44	1	33 58	10
03 38	35	10 35	24	16 48	0	21 59	6	28 45	1	34 03	5
03 45	7	10 35	0	16 57	9	22 00	1	28 49	4	34 13	10
03 51	6	10 41	6	17 06	9	22 23	23	28 49	0	34 15	2
03 53	2	10 56	15	17 08	2	22 27	4	28 50	1	34 17	2
03 57	4	10 59	3	17 19	11	22 27	0	28 52	2	34 36	19

Time		Time		Time		Time		Time		Time	
03 58	1	11 02	3	17 27	8	22 30	3	28 53	1	34 56	20
04 02	4	11 09	7	17 28	1	22 39	9	28 53	0	35 02	6
04 03	1	11 09	0	17 37	9	22 51	12	28 54	1	35 06	4
04 18	15	11 19	10	17 40	3	22 55	4	28 55	1	35 20	14
04 28	10	11 23	4	17 41	1	23 02	7	28 57	2	35 38	18
04 29	1	11 26	3	17 42	1	23 06	4	28 58	1	35 57	19
04 30	1	11 41	15	17 46	4	23 34	28	28 59	1	36 01	4
04 34	4	11 43	2	17 56	10	23 38	4	29 05	6	36 26	25
05 05	31	11 46	3	18 11	15	23 43	5	29 20	15	36 40	14
05 05	0	11 53	7	18 22	11	23 57	14	29 46	26	36 46	6
05 08	3	11 55	2	18 25	3	24 07	10	30 06	20	36 51	5
05 26	18	12 05	10	18 27	2	24 08	1	30 09	3	36 54	3
05 47	21	12 11	6	18 28	1	24 14	6	30 10	1	36 58	4
05 48	1	12 27	16	18 31	3	24 18	4	30 15	5		
05 50	2	12 33	6	18 33	2	24 41	23	30 24	9		
05 57	7	12 51	18	18 34	1	24 42	1	30 28	4		
06 00	3	12 53	2	18 43	9	24 51	9	30 41	13		
06 04	4	12 56	3	18 49	6	24 55	4	30 42	1		
06 24	20	13 16	20	18 55	6	25 11	16	30 42	0		
06 34	10	13 23	7	18 56	1	25 13	2	30 43	1		
06 39	5	13 30	7	19 18	22	25 31	18	30 45	2		
06 45	6	13 39	9	19 19	1	25 38	7	30 52	7		
07 07	22	13 45	6	19 26	7	25 42	4	30 55	3		
07 08	1	14 05	20	19 29	3	25 50	8	31 26	31		
07 27	19	14 08	3	19 42	13	25 55	5	31 26	0		

Swedish registered. (Note that Elsinore in Denmark is connected to Hälsingborg in Sweden by ferry.)

Example 7. The Rutherford-Geiger experiment given in §10 in fact consists of 4 sub-experiments as shown in Table 15.

Table 15 The Rutherford-Geiger experiment

k	0	1	2	3	4	5	6	7	8	9	10	11	12	13	14	≥15
$N_k(I)$	15	56	106	152	170	122	88	50	17	12	3	0	0	1	0	0
$N_k(II)$	17	39	88	116	120	98	63	37	4	9	4	1	0	0	0	0
$N_k(III)$	15	56	97	139	118	96	60	26	18	3	3	1	0	0	0	0
$N_k(IV)$	10	52	92	118	124	92	62	26	6	3	0	2	0	0	1	0

Example 8. (see [8], p. 781). Table 16 shows, for different values of t_1 and t_2, the number of outbreaks of ionisation due to cosmic radiation, where an outbreak follows τ seconds after the outbreak last observed, for a value of τ between t_1 and t_2. The total time of observation was 35 hours.

Table 16

t_1(sec.)	t_2(sec.)	number
0	50	22
50	100	17
100	200	26
200	500	68
500	1000	47
1000	2000	31
2000	5000	2
5000	∞	0

Example 9. (from [1]). Table 17 shows the arrival and inter arrival times for customers who joined a queue for a cashier in a department store. The data was collected by photographing with time-marked film. The unit of time was 1 second.

Example 10. Tables 18 and 19 show the numbers of registered cases of illness in a certain period in a certain region, for jaundice and leukemia respectively. The numbers in the left-hand columns show the number of days elapsed since the start of observations (arrival times) and the numbers in the right-hand columns show the corresponding waiting times. In the original data, for which the University of London has the copyright, the place of each case of illness was also given.

Table 17 Customers' arrival times

	31		4		36		14		12		13
31		505		1379		2189		2856		3380	
	13		22		131		4		0		94
44		527		1510		2193		2856		3474	
	7		52		1		26		71		37
51		579		1511		2219		2927		3511	
	45		22		70		71		0		31
96		601		1581		2290		2927		3542	
	24		20		4		10		0		5
120		621		1585		2300		2927		3547	
	51		83		33		10		2		16
171		704		1618		2310		2929		3563	
	46		44		22		28		17		38
217		748		1640		2338		2946		3601	
	7		12		16		1		10		9
224		760		1656		2339		2956		3610	
	8		3		20		54		73		27
232		763		1676		2392		3029		3637	
	1		39		19		4		12		9
233		802		1695		2397		3041		3646	
	4		53		3		53		31		23
237		855		1698		2450		3072		3669	
	22		12		24		160		5		64
259		867		1722		2610		3077		3733	
	28		10		32		27		27		34
287		877		1754		2637		3104		3767	
	12		18		90		10		27		6
299		895		1844		2647		3131		3773	
	6		63		61		11		2		0
305		958		1905		2658		3133		3773	
	65		62		77		7		0		22
370		1020		1982		2665		3133		3795	
	16		26		6		34		0		57
386		1046		1988		2699		3133		3852	
	9		25		105		1		99		23
395		1071		2093		2700		3232		3875	
	12		47		0		38		10		20
407		1118		2093		2738		3242		3895	
	27		26		13		3		30		41
434		1144		2106		2741		3272		3936	
	2		105		2		4		7		12
436		1249		2108		2745		3279		3948	
	44		27		27		37		1		36
480		1276		2135		2782		3280		3984	
	1		1		17		6		10		2
481		1277		2152		2788		3290		3986	
	10		42		0		31		28		10
491		1319		2152		2819		3318		3996	
	10		24		23		25		9		49
501		1343		2175		2844		3327		4045	
									40		
								3367			

Table 18 Registered cases of jaundice in Dorset, May 1938–November 1944

8	8	733	493	1821	2	2245	11
37	29	739	6	1828	7	2255	10
141	104	750	11	1829	1	2255	0
148	7	754	4	1832	3	2257	2
149	1	757	3	1843	11	2264	7
149	0	764	7	1928	85	2265	1
151	2	781	17	1932	4	2320	55
152	1	783	2	1984	52	2340	20
157	5	807	24	1986	2	2342	22
169	12	820	13	1990	4	2348	6
171	2	968	148	1990	0	2382	34
171	0	979	11	1993	3		
172	1	1139	160	2019	26		
173	1	1567	428	2121	102		
184	11	1697	130	2155	34		
185	1	1700	3	2174	19		
192	7	1706	6	2196	22		
197	5	1738	32	2199	3		
202	5	1742	4	2206	7		
202	0	1742	0	2208	2		
202	0	1774	32	2222	14		
207	5	1789	15	2227	5		
211	4	1789	0	2229	2		
223	12	1810	21	2234	5		
240	17	1819	9	2234	0		

Table 19 Registered cases of leukemia in Cornwall, April 1947-January 1962

	26		64		280		3
26		1813		2966		4079	
	86		8		6		131
112		1821		2972		4210	
	252		40		133		103
364		1861		3105		4313	
	5		3		26		2
369		1864		3131		4315	
	94		3		38		103
463		1867		3169		4418	
	70		6		5		65
533		1873		3174		4483	
	35		23		29		53
568		1896		3203		4536	
	222		2		8		46
790		1898		3211		4582	
	64		29		96		19
854		1927		3307		4601	
	0		24		17		82
854		1951		3324		4683	
	21		0		8		1
875		1951		3332		4684	
	24		9		17		33
899		1960		3349		4717	
	24		107		51		3
923		2067		3400		4720	
	295		67		54		114
1218		2134		3451		4834	
	167		4		2		5
1385		2138		3456		4839	
	17		123		76		196
1402		2261		3532		5035	
	20		17		26		77
1422		2278		3558		5112	
	98		95		28		5
1520		2373		3586		5117	
	102		0		97		6
1622		2373		3683		5123	
	53		103		119		176
1675		2476		3802		5299	
	65		119		70		107
1740		2595		3872		5406	
	9		91		204		
1749		2686		4076			

Example 11. Do the durations of telephone conversations follow an exponential distribution? Erlang wrote a paper in 1920 based on data from 2461 telephone conversations in 1916 through the Copenhagen main exchange, classified by duration at 10 second intervals. The corresponding histogram is given in Figure 22; the numbers on the vertical axis give the percentage of the 2461 conversations.

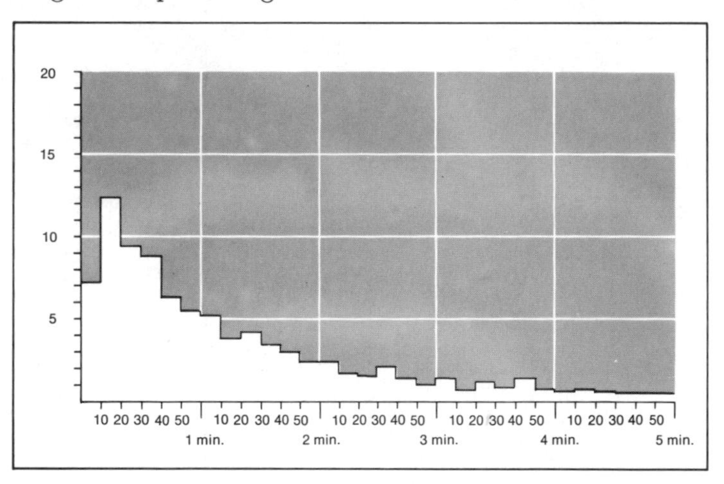

Fig. 22. Histogram of durations of telephone conversations.

Example 12. Table 20 shows statistics of the number of accidents among 647 female workers at a shell factory over a 5-week period.

Table 20

number of accidents	number of workers
0	447
1	132
2	42
3	21
4	3
5	2
≥ 6	0

Remark: The Poisson approximation gives only a mediocre result. The data stems from a 1920 article of Greenwood and Yule, and it was precisely their intention to show that deviations from the pure Poisson model can often occur. They also wanted to point out that deviations can be "explained" by working with mixtures of Poisson distributions. In the present case this corresponds to workers having different tendencies to become exposed to accidents. Here we cannot go into the mixed

Poisson distributions, but refer to Hald's book [11], §20.9. As a simplification one could perhaps consider a model in which a certain percentage of workers have one "accident tendency" and the rest another accident tendency (see Exercise 33, (iv)).

Example 13.[23] On 28 October 1981 the Soviet submarine U137 was discovered off the Blekinges archipelago, not far from Karlskrona, Sweden. The submarine had put itself at rest on one of those rocks which, also to Danish sailors, must be said to be rather well known.

On November 5 Prime Minister Fälldin announced that the submarine carried uranium 238 in such a quantity that it could only be attributed to nuclear weapons. The announcement was based on measurements carried out by the defence research institution FOA.

The decay of uranium 238 sets off a long series of processes. What is involved is not a pure cascade process (Exercises 40-42), but a *cascade process with branching*, since several of the nuclei in the radioactive series can decay in two or more different ways, with certain associated probabilities. Since we find it natural to apply a deterministic model, we shall say that a certain percentage of decays take such and such a form.

The most energetic radiation emitted in the decay of uranium 238 is γ-radiation with an energy of 1.001 MeV. This is precisely the radiation that interests us. The part of the radioactive series we need to know can be summarised in the following diagram:

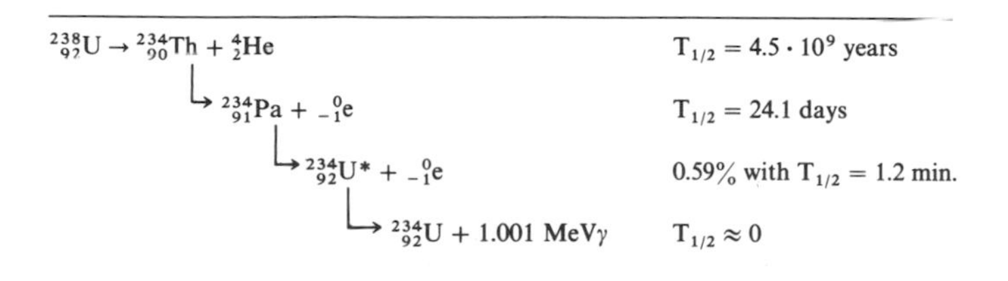

[23] The sources of this example are FOA Tidningen, no. 4, 1981, a broadcast of Swedish TV, 26 February 1982, and information obtained from Ragnar Hellborg, Lund University and Ulf Jacobsen, RISØ.

Thus it is only 0.59% of the protactinium atoms which decay to the excited state of uranium 234, and the excited uranium atoms decay immediately to the ground state with the emission of a high energy γ-quantum.

We give references below to all the relevant information that is generally accessible and concerns the measurement of the γ-radiation in question. The problem is to give an estimate of the amount of uranium 238 in the source. In the FOA's report one can read:

After more than four hours' measurement, one could confirm not only that ^{238}U had been identified with 100% certainty, but also that inside the submarine at the point measured there were large quantities of this material, to be counted in kilograms.

Measurement of γ-radiation

Registered intensity of $1 MeV$ *γ-radiation:*	500 counts per hour.
Detector area:	20 cm^2.
Detector efficiency for $1 MeV$ *γ-radiation:*	1%.
Distance of detector from source:	1.5 m.
Thickness of iron which the radiation has to penetrate (torpedo tube + hull):	$\approx$ 24 mm.
Attenuation constant for passage of $1 MeV$ *γ-radiation through iron:*	0.5 cm^{-1}.

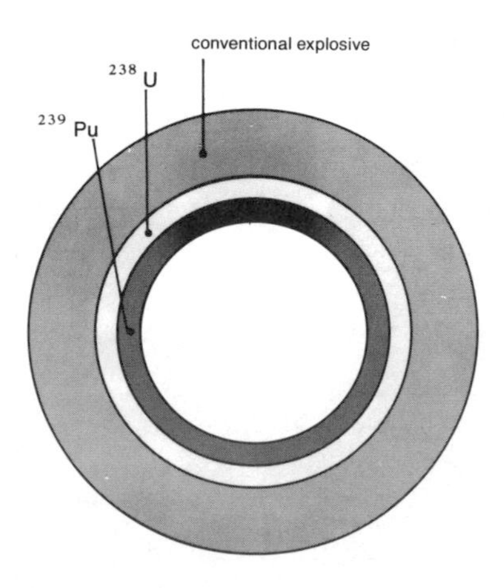

Fig. 23. Principal sketch of a nuclear charge of implosion type.

Unfortunately it has been impossible to obtain information on the geometry of the source and the attenuation in its immediate neighbourhood. The FOA paper merely pointed to an existing type, see Figure 23. The attenuation properties of this type are not easily estimated.

Those who wish may try to work out whether it was at all reasonable to claim there was a large quantity of uranium 238 in the source, to be counted in kilograms.

Example 14.[24] Three important areas of age determination based on the theory of radioactive decay concern organic material, volcanic rocks and the earth.

Even in the case of a simple mother-daughter reaction, where the daughter atoms are stable, age determination rarely goes as simply as was indicated in Exercise 43. One reason for this is that often a certain number of daughter atoms are already present in the material at the time of its formation. One can get over this difficulty by taking measurements from several equally old samples (see below). Another possibility, which can be used in determining the age of the earth, is to use the special circumstances of meteorites (particularly the insolubility of uranium in iron) to obtain information on the proportions of isotopes in the earth at the time of its origin.

We shall look at an example of age determination of some rocks from a locality in Ontario. The method is the often-used *rubidium-strontium method*, which uses the process

$$^{87}_{37}\mathrm{Rb} \longrightarrow {}^{87}_{38}\mathrm{Sr} + {}^{0}_{-1}e, \qquad T_{1/2} = 5 \cdot 10^{10} \text{ years.}$$

The daughter nucleus is stable.

For technical reasons one does not measure absolute numbers of rubidium and strontium atoms, but rather their values relative to the proportion of strontium 86 isotope. Since strontium 86 is not a decay product of any known radioactive series, one can assume the concentration of strontium 86 in a rock to be constant. In addition, one can assume the ratio between strontium 87 and strontium 86 at the time of formation to be the same for rocks in the same locality.

Let y denote the ratio $^{87}\mathrm{Sr}/^{86}\mathrm{Sr}$ and x the ratio $^{87}\mathrm{Rb}/^{86}\mathrm{Sr}$, both ratios measured at the present time. Figure 24 shows together the values of x and y corresponding to 6 rocks from the same locality.

[24] The example is from Faure's book [9] (§6.2). We refer to it for a basic discussion of this example together with a treatment of the whole fascinating circle of related ideas.

By starting from the equation in Exercise 44, one comes to see that the rocks must be around 1.7 billion years old. Think it over!

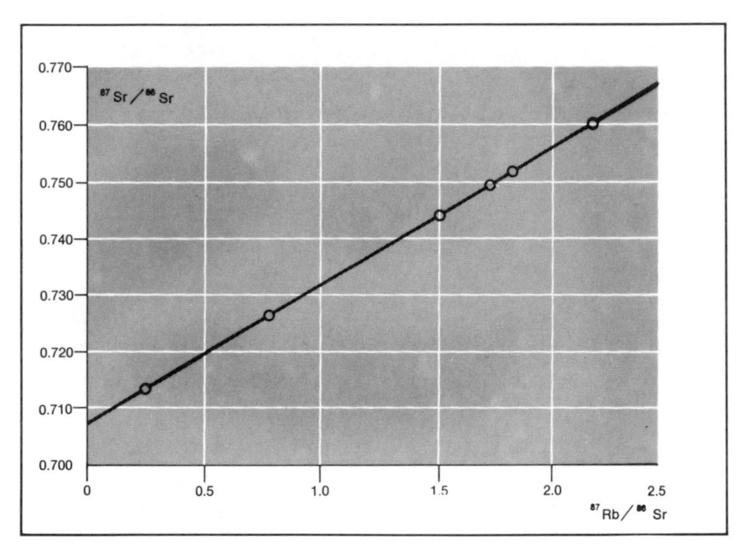

Fig. 24.

Example 15. It is possible to carry out school experiments with short-lived isotopes. We shall look at the process

$$
\begin{aligned}
{}^{137}_{55}\mathrm{Cs} \longrightarrow {}^{137}_{56}\mathrm{Ba}^* + {}^{0}_{-1}e \qquad & T_{1/2} = 30 \text{ years} \\
\hookrightarrow {}^{137}_{56}\mathrm{Ba} + 662 \text{ keV } \gamma \quad & T_{1/2} = 2.6 \text{ min.}
\end{aligned}
$$

With a special arrangement involving a "minigenerator," an experiment can be carried out in such a way that at the start (time $t = 0$) there are only Ba* atoms (excited Barium atoms). With the help of a counter one can record the γ-radiation arising from decay to the ground state of barium. The activity of Ba* is then equal at all times to the intensity of this γ-radiation (cf. Exercise 39). The intensity is measured at time t by carrying out an integration over the time interval $[t - T, t]$, where T is a suitably chosen constant. T must not be so small as to give a wholly random variation, nor so large that the measured value is unrepresentative of the situation at time t. The integration itself is carried out with the help of suitably connected electronic equipment.

The result of an experiment is given in Figure 25. Notice that one can clearly recognise the stochastic nature of the phenomenon, while at

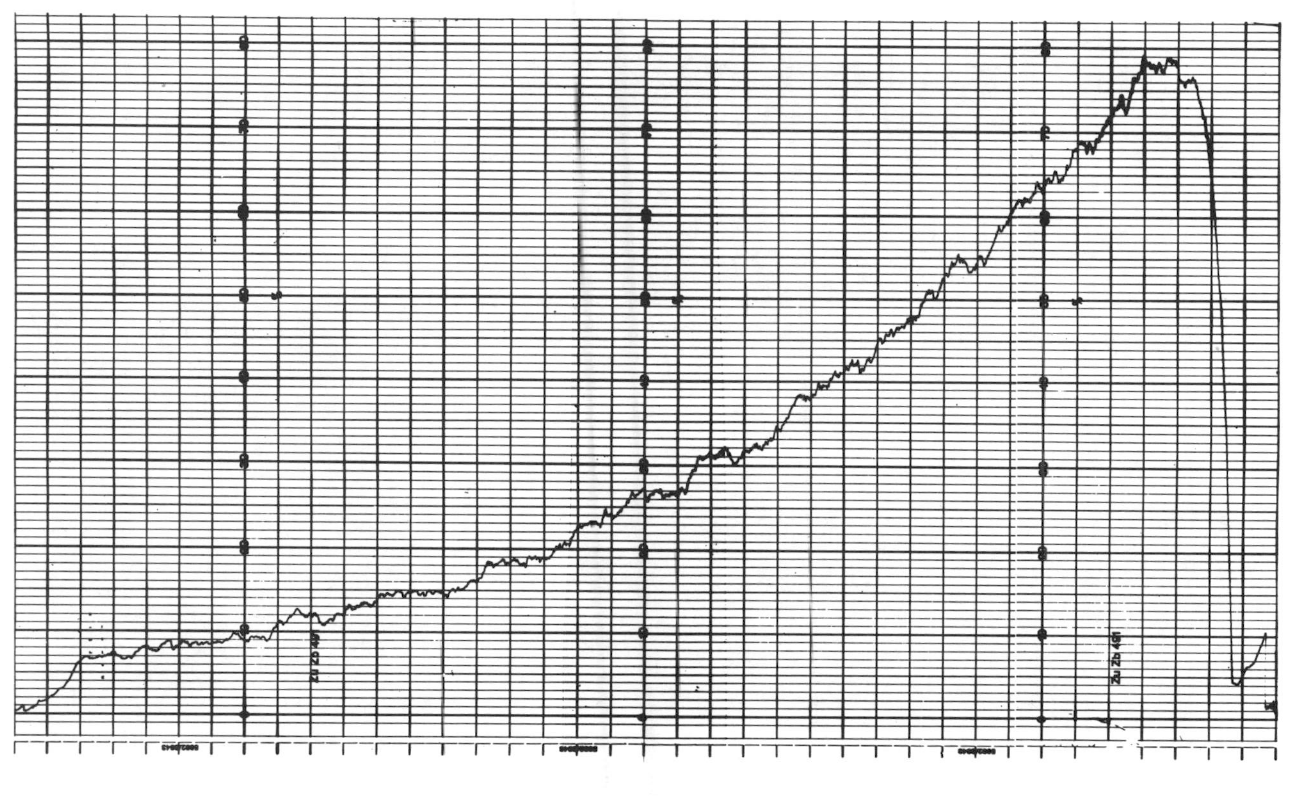

Fig. 25. Experiment with the Cesium-Barium minigenerator. The automatic writer moves from right towards left with a speed of 3 subintervals per minute. The numbers on the vertical scale denote counts per second. (Experiment by Malte Olsen).

the same time there is an obvious approximation to the given activity curve for Ba* by a deterministic curve.

In treating this example, one should check whether the given half-life of 2.6 minutes is reasonable (see also Exercise 42).

Example 16. The results of the World Cup in soccer in 1986 are tabulated below (the result of each match is the result after ordinary playing time).

First round

Bulgaria–Italy	1–1	Canada–France	0–1	Scotland–Denmark	0–1
Argentina–South Korea	3–1	Soviet Union–Hungary	6–0	Uruguay–West Germany	1–1
Italy–Argentina	1–1	France–Soviet Union	1–1	West Germany–Scotland	2–1
Bulgaria–South Korea	1–1	Hungary–Canada	2–0	Uruguay–Denmark	1–6
South Korea–Italy	2–3	France–Hungary	3–0	Denmark–West Germany	2–0
Argentina–Bulgaria	2–0	Soviet Union–Canada	2–0	Scotland–Uruguay	0–0
Mexico–Belgium	2–1	Spain–Brasil	0–1	Poland–Maroc	0–0
Paraguay–Iraq	1–0	Northern Ireland–Algeria	1–1	Portugal–England	1–0
Mexico–Paraguay	1–1	Brasil–Algeria	1–0	England–Maroc	0–0
Iraq–Belgium	1–2	Northern Ireland–Spain	1–2	Poland–Portugal	1–0
Paraguay–Belgium	2–2	Brasil–Northern Ireland	3–0	Portugal–Maroc	1–3
Iraq–Mexico	0–1	Spain–Algeria	3–0	England–Poland	3–0

1/8-finals		**quarter-finals**		**semi-finals**	
Soviet Union–Belgium	2–2	Brasil–France	1–1	France–West Germany	0–2
Mexico–Bulgaria	2–0	West Germany–Mexico	0–0	Argentina–Belgium	2–0
Brasil–Poland	4–0	Argentina–England	2–1		
Argentina–Uruguay	1–0	Spain–Belgium	1–1	**Match about third placement**	
Italy–France	0–2			France–Belgium	2–2
Maroc–West Germany	0–1				
England–Paraguay	3–0				
Denmark–Spain	1–5				

final

Argentina–West Germany 3–2

To provide a comparison with the World Cups of '74, '78 and '82, Table 21 show thw number of goals scored; it also shows the number of games in which the total number of goals was 0, 1, 2, etc.

Table 21 Soccer World Championships

	WC-74	WC-78	WC-82	WC-86
0	5	6	7	4
1	6	7	9	10
2	11	3	12	17
3	8	8	7	10
4	2	8	6	5
5	3	3	8	2
6	1	3	1	3
7	1	0	1	1
8	0	0	0	0
9	1	0	0	0
10	0	0	0	0
11	0	0	1	0

Example 17.[25] Patients with multiple sclerosis are often observed to have a condition that is harmless in itself, making itself known by faint stripes around the veins in the retina of the eye. We call this condition PR (*periphlebitis retinae*). PR occurs without any apparent reason, and it vanishes after some time, likewise without any apparent cause. Neither the appearance nor disappearance of PR is noticed by the patients; it is first uncovered by a clinical investigation, which, by the way, is also likely to be clouded by some uncertainty.

There is no case reported in the literature in which PR has appeared more than once in the same patient. It would be interesting to know whether this is actually the case, since it might give medical science some information about the character of multiple sclerosis, the cause of which is still unknown.

The source material yields the following information, presented here in simplified form.

The mean duration of multiple sclerosis is estimated to be 30 years.

Of 3021 patients whose case histories are reported in the literature, 346 were found to have PR at the time of examination. (because of difficulties in the examination itself, some cases of PR were probably not registered).

Table 22

remaining duration r, approximate value in months	number of patients
2	3
3	2
5	1
6	4
12	1
25	1
$1 \leq r \leq 12$	1
$1 \leq r \leq 16$	1
$12 \leq r \leq 20$	1
$1\frac{1}{2} \leq r$	1
$3 \leq r$	1
$4 \leq r$	1
$5 \leq r$	2
$10 \leq r$	1
$11 \leq r$	2

In very few cases, 23 altogether, an attempt was made to determine the remaining duration of PR, i.e. the interval between the time PR

[25]The example is from a paper by Tine Engell and Per Kragh Andersen in *Acta Neurologica Scandinavica*, 1982.

was discovered and the time it vanished. The results are given in Table 22. The authors of the paper conclude that one may assume that PR can appear several times in a patient's life (about $3\frac{1}{2}$ times on the average). In this example the problem is to make up one's mind about the reasonableness of this conclusion. One can set up a model for the distribution of the total duration of PR and try to estimate the mean duration from the given data on the remaining duration (the footnote to Exercise 22 may be of help here).

Example 18.[26] Bacteriophages are a type of virus that is parasitic to bacteria. They reproduce themselves by crowding into cells of bacteria. When a bacterium is infected, a certain time elapses, during which new bacteriophages are formed. Eventually the cell walls of the bacterium burst, and the newly formed bacteriophages break free. One says that the bacterium *lyses.* The formation process can then be repeated.

There have been numerous experiments to confirm this theory and elaborate on it, the first being presented by F. d'Herelle in 1917. An experiment in 1938 by E.L. Ellis and H. Delbrück, two of the pioneers in the field, was concerned especially with estimating how many bacteriophages broke free from a single infected bacterium.

In the experiment, a *coli* bacteria culture was added to a very dilute solution with bacteriophages. After 10 minutes had elapsed, one could assume that almost all bacteriophages had entered the cells of the bacteria. A further dilution was then carried out and 40 samples were taken in small test tubes, each containing 0.05 ml. The dilution was to ensure that the samples contained very few, perhaps no, cells with bacteriophages. After an incubation (hatching) time of 200 minutes, the total number of bacteriophages in each test tube was counted by a standard technique, in which the solution was poured over an agar plate, specially prepared with indicator bacteria, after which a count was made of the number of *plaques* formed (visible light regions caused by attacks from bacteriophages). Each plaque corresponds to one bacteriophage. Table 23 shows the plaque counts obtained in the experiment.

The reader who takes advantage of the normal distribution will perhaps be able to set up a reasonable model of all the data, but it is not necessary to do this in order to extract useful information on the number of bacteriophages produced per infected cell of bacteria.

---

[26]See G.S. Stent: *Molecular Biology of Bacterial Virus*, W.H. Freeman and Co., 1963, especially chapter 4. Professor Erik Bahn has assisted with this example.

It should be remarked that the very small positive counts which appear could be due to the presence of bacteriophages that had not become attached to bacteria cells at the time the 40 samples were collected.

Table 23 Plaque counts from 40 samples

0	26	123	31	0	0	45	0
130	0	83	0	0	48	0	190
0	0	0	0	0	1	0	0
0	0	9	5	53	0	0	9
58	0	0	0	0	72	0	0

Programs

The programs are written in GW-BASIC from Microsoft and tested on both a portable computer, the ai-PC16 from ai Electronics, and an IBM Personal System/2. Only the more primitive BASIC-commands have been used.

When writing a program, one often aims at making it numericly robust, fast, compact, and user-friendly, but none of these criteria have been applied to these programs to any great extent. The point we want to make is that the reader should understand how the programs function so that he can adapt them to his own taste and equipment. For example, one can easily control the input of data from special data files and refine the printouts via PRINT USING commands. But the programs can be used as they are, if the reader wishes.

The individual program parts P1-P8, to which reference is made in the text, have been assembled into one program by the control program P0. Each one of the parts P1-P8 may be implemented as an independent program.

At the end of this chapter, the reader will find a printout off all the program parts P0-P8 as well as a printout from a run of the program. The run relates to the key example discussed in Chapter 16.

P1. Storage of data.

This program is used to store data depending on an integer valued parameter and to calculate certain numbers (see below). Basically, the program consists of three parts: input (l. 1000-1090), control (l. 1140-1190) and corrections (l. 1200-1250).

Assume that you wish to store the numbers $x(k); k = k_1, \cdots, k_2$. You first key in the extreme parameter values k_1 and k_2. They are stored in E and F, respectively. The program sets $G = F - E$. Because many of the datasets we study contain a time constant, you are asked to key in this constant (stored in T). For instance, in the Rutherford-Geiger experiment, you should set $T = 7.5$. If no time constant is associated to the dataset, the program itself will set $T = 1$ (this is the default value).

The numbers $x(k)$ are keyed in one by one as the corresponding parameter value k is displayed on the screen. Storage of $x(k)$ is in $A(k - k_1); k = k_1, \cdots, k_2$. The necessary storage space is created in line 1030.

When the keying is finished, the variables U, V and W will have the values $\sum_0^G A(N)$, $\sum_0^G (N+E)A(N)$ and $\sum_0^G (N+E)^2 A(N)$, respectively.

When activating the control, the stored data are displayed as a vector $A(0), A(1), \cdots, A(G)$. If any of the data have been entered incorrectly, you can activate the correction part of the program. When all corrections have been carried out, $A(N) = x(N+E); N = 0, 1, \cdots, G$; and U, V and W have the values $\sum x(k), \sum kx(k)$ and $\sum k^2 x(k)$,respectively. By definition, these numbers are the *moments* associated with the $x(k)'$s of *orders* 0, 1 and 2, respectively. These moments are displayed when controlling the data together with the observed intensity per unit time (assuming that the data have an interpretation as by the Rutherford-Geiger experiment).

Here is an indication of numerical problems which may arise: If you key in some data, make some corrections, and then correct back to the original data, you should of course, get back the original moments. However, this is not certain! Consider! As a very general guide concerning numerical calculations, one can say that it is wise to work with numbers and operators which do not vary to the extreme from each other.

P2. Tabulation of Poisson - and Binomial distributions.

This program leads to tabulation of the individual probabilities p_k as well as the accumulated probabilities $p_0 + p_1 + \cdots + p_k$, possibly multiplied by a constant (stored in M).

The program requests that you enter the value 1 if you want to deal with the Poisson distribution and 2 if you desire the binomial distribution. Furthermore, you are asked to enter a value (stored in D) giving the lower limit of values of Mp_k which results in display or printout.

The numbers Mp_k are stored in A. At first, A is assigned the value Mp_0, then the value Mp_1, and so on. The corresponding accumulated numbers $M(p_0 + \cdots + p_k)$ are stored in B. If $Mp_k \geq D$, then the value of k (stored in K) is displayed together with the values of Mp_k and $M(p_0 + \cdots + p_k)$.

The essential basis for the calculations prescribed by the program is the recursion formulae:

$$p_k = p_{k-1}\frac{\lambda}{k}; \qquad k = 1, 2, \cdots$$

(Poisson distribution),

$$p_k = p_{k-1}\frac{k}{n-k+1}\ \frac{p}{1-p}; \qquad k = 1, 2, \cdots, n$$

(binomial distibution).

P3. Calculation of dead time by speciel formula.

The formula in question is (42). Together with (41), this formula is used to determine λ and h. To run this program you first have to run P1.

The essential part of P3 is contained in the subroutine in lines 3140-3190. Here, the difference J between the right-hand side and the left-hand side in (42) is calculated for a given value of h (stored in H).

To execute the program, one starts by giving two guesses for the value of h (the guesses are stored in A and B). One of the values could be 0. The subroutine will then give you the corresponding differences (stored in P and R). To obtain the next guess for the value of h (temporarily stored in C), the program simply uses linear interpolation).

The process is iterated and does not stop until both differences corresponding to the two current values of h are 0 (however, iterations may be stopped by the user). When this happens, the two values found for h and λ (stored in H and L, respectively) are displayed.

During execution, the two values for h with corresponding differences are displayed for each iteration. You can therefore follow the process in detail and, for instance, detect any numerical instability. This also offers the user the opportunity to break off the execution in case you are satisfied with the attained degree of accuracy.

The program may be adapted to find the zeroes of arbitrary functions by replacing the subroutine by a routine which calculates the corresponding function value for a given value of H.

It should be noted that it is quite likely that the program will continue running unless you explicitly stop it. For instance, this happens if there is no value of h for which the subroutine will return the difference 0 (even though this situation is excluded theoretically by exact calculation).

P4. Tabulation of the modified Poisson distribution.

These calculations utilize formula (40). As input, the values of T, H and L (t, h and λ of (40)) are demanded. If these quantities have been determined by running the programs P1 and P3, the values obtained can be used directly.

To avoid unnecessary calculations, you are asked to enter the smallest N and the largest N for which you want to have a display of probabilities and accumulated probabilities. Furthermore, you are given the option of having these probabilities multiplied by a number M before display. This is convenient for the calculation of expected values; in case P1 has been executed, one may set $M = U$.

Concerning the organization of the calculations, we remark that the current accumulated probabilities (multiplied by M) are stored in S, while the corresponding previous values are stored in Q. For a given N, the subroutine in lines 4180-4240 calculates the associated accumulated probabilities multiplied by M.

P5. Calculation of minus the log-likelihoodfunction for the modified Poisson distribution.

For given values of h and λ (stored in H and L), this program calculates minus the natural logarithm of the likelihoodfunction of the modified Poisson distribution with parameters λ, h and T and corresponding to a sample which is assumed to have been keyed in utilizing program P1. The subroutine in lines 5180-5240 is by and large identical to the subroutine of P4.

P6. Determination of dead time by the two source method.

This program is based on the formula (52) from Exercise 33. The dead time found is stored in K.

P7. χ^2-test for goodness of fit to the Poisson distribution.

It is assumed that observed data have been keyed in utilizing program P1. Normally, the test should be carried out for the Poisson distribution with parameter λ equal to L calculated from P1. If not instructed otherwise, the program assumes that $\lambda = L$ and that the grouping limit is 5.

The grouping to be used is determined from L and X (this is where the grouping limit is stored) without utilizing the observed data, c.f. the discussion following Theorem 4 in Chapter 17.

To understand the structure of the program it suffices to know the meaning assigned to the various storage cells. K is a parameter assuming the values $0, 1, 2, \cdots$. The expected number of observations assuming the value K is stored in A while the expected number of observations assuming a value $\leq K$ is stored in B. In Q we store the accumulated expected number of observations corresponding to all groups already handled. In N we store the actual number of observations in the group being treated. The various contributions to the discrepancy which are calculated during execution are accumulated in D so that D at last (1.7150) has the correct value. Finally, the number of groups in the grouping are counted in J.

As soon as a group has been formed, the following quantities are displayed (l. 7090): The largest number in the group, the number of actual observations in the group, the expected number of observations in the group, and the accumulated expected number og observations corresponding to all groups created so far. The last group consists of all numbers larger than a certain number. This is indicated by display of the sign ˆ. The display is terminated by giving the discrepancy and the number of groups in the grouping.

If you employ this program, the program P2 is made superfluous to some extent.

P8. χ^2-test for goodness of fit for the modified Poisson distribution.

In principle, this program is analogous to the previous one. It demands previous entry of data via program P1. The dead time and the intensity are also required as inputs. These quantities could, for instance, have been determined by prior execution of the program P3 (in which case the values obtained from it will be employed automatically). Finally, the grouping limit must be given (the value 5 is suggested) and, in order to save time, a value of K (with the same significance as in P7) for which you want to start the calculations is also requested. This value, stored in C, should be chosen as large as possible, but it must hold that $UP_{\lambda,h}(N_{\text{reg}}(t) \leq C)$ is below the grouping limit.

When, for a given value of $K (\geq 1)$, the subroutine in l. 8240-8280 is called, the probability $P_{\lambda,h}(N_{\text{reg}}(t) \leq K)$ is calculated and stored in B.

Contrary to program P7, the present program works with the last accumulated value $P_{\lambda,h}(N_{\text{reg}}(t) \le K-1)$, stored in P, in addition to the quantity B.

If you make use of this program, the program P4 is made superfluous to a certain extent.

Programs P0-P8

P0

```
10  O$="*****************":PRINT O$+O$;"  BEGIN  ";O$+O$
20  INPUT "Choice of Program-Section (1/2/3/4/5/6/7/8) or End<ENTER>:",I$
30  IF I$="" GOTO 130
40  PRINT O$+O$;" Begin P";I$;O$+O$
50  IF I$="1" GOTO 1000
60  IF I$="2" GOTO 2000
70  IF I$="3" GOTO 3000
80  IF I$="4" GOTO 4000
90  IF I$="5" GOTO 5000
100 IF I$="6" GOTO 6000
110 IF I$="7" GOTO 7000
120 IF I$="8" GOTO 8000
130 IF I$="" THEN INPUT "sure<ENTER>/ escape(any other key):",I$
140 IF I$="" THEN PRINT O$+O$;"   END   ";O$+O$:END
150 GOTO 20
```

P1

```
1000 INPUT "             min# ";E
1010 INPUT "             max# ";F
1020 INPUT "time (default=1) ";T:IF T=0 THEN T=1
1030 G=F-E:DIM A(G)
1040 U=0:V=0:W=0
1050    FOR N=0 TO G
1060    R=N+E:PRINT R;:INPUT A(N)
1070    Q=A(N):U=U+Q:V=V+R*Q:W=W+R^2*Q
1080    NEXT N
1090 L=V/U/T:PRINT
1100 INPUT "               Control(1)/ Correction(2)/ End P1<ENTER>:",I
1110 PRINT:IF I=1 GOTO 1140
1120 IF I=2 GOTO 1200
1130 PRINT O$+O$;" End P1 ";O$+O$:GOTO 20
1140    FOR N=0 TO G
1150    PRINT A(N);:IF N<G THEN PRINT ",";
1160    NEXT N: PRINT
1170 PRINT "Moments (0/1/2): ";U;V;W
1180 PRINT "Observed Intensity (# events per unit time):";L
1190 PRINT:GOTO 1100
1200 INPUT "Correction of No.: ",R
1210 INPUT "      true value : ",Q:PRINT
1220 IF (R-E)*(F-R)<0 GOTO 1200
1230 N=R-E:D=Q-A(N):A(N)=Q
1240 U=U+D:V=V+R*D:W=W+R^2*D
1250 GOTO 1090
```

P2

```
2000 INPUT "Poisson Distribution(1)/ Binomial Distribution(2):",X:PRINT
2010 IF X<>1 THEN IF X<>2 GOTO 2210
2020 IF X=1 THEN PRINT "If lambda =";L;"<ENTER>, else new value of lambda";
2030 IF X=1 THEN INPUT ":",I:IF I<>0 THEN L=I
2040 IF X=2 THEN INPUT "n ";N:INPUT "p ";P
2050 INPUT "                    Multiplier (default=1) ";M:IF M=0 THEN M=1
2060 INPUT " Minimal value (prob.* multiplier) by display ";D
2070 INPUT "          Max. value of k by display ";I:PRINT
2080 IF X=1 THEN A=M/EXP(L)
2090 IF X=2 THEN A=M*(1-P)^N
2100 IF X=1 THEN PRINT "POISSON DISTRIBUTION with lambda =";L;
```

```
2110 IF X=2 THEN PRINT "BINOMIAL DISTRIBUTION with n =";N;",";"p =";P;
2120 PRINT "  (Multiplier =";M;"):":PRINT
2130 K=0:B=A:J=0
2140 IF A>=D THEN PRINT K;": ";A;" ";B:J=J+1:IF J=18 GOTO 2190
2150 K=K+1:IF K>I GOTO 2210
2160 IF X=1 THEN A=A*L/K
2170 IF X=2 THEN A=A*(N-K+1)/K*P/(1-P)
2180 B=B+A:GOTO 2140
2190 INPUT "continue<ENTER>/ End(any other key):",I$
2200 IF I$="" THEN J=0:GOTO 2150
2210 PRINT O$+O$;"  End P2 ";O$+O$:GOTO 20
```

P3

```
3000 IF V^2/U<W-V THEN PRINT "h=0, lambda=";V/U/T:GOTO 3130
3010 INPUT " first guess ";A
3020 INPUT "second guess ";B:PRINT
3030 H=A:GOSUB 3140:P=J:Z=1
3040 H=B:GOSUB 3140:R=J
3050 PRINT "Iteration No.";Z;":";A;P:PRINT "             ";B;R:Z=Z+1
3060 IF INT(Z/9)=Z/9 THEN INPUT "continue<ENTER>/ break (any other key)",I$
3070 IF INT(Z/9)=Z/9 THEN IF I$<>"" GOTO 3130
3080 IF P=R THEN C=(A+B)/2:IF P=0 GOTO 3120
3090 IF P<>R THEN C=A+P*(B-A)/(P-R)
3100 IF ABS(C-A)<=ABS(C-B) THEN B=C:GOTO 3040
3110 A=B:P=R:B=C:GOTO 3040
3120 L=1/(U*T/V-C):H=C:PRINT:PRINT "h=";H,"lambda=";L
3130 PRINT O$+O$;"  End P3 ";O$+O$:GOTO 20
3140    S=0
3150       FOR K=0 TO G
3160       S=S+(K+E)*(K+E-1)*A(K)/(T-(K+E)*H)
3170       NEXT K
3180    J=V/(U*T/V-H)-S
3190    RETURN
```

P4

```
4000 PRINT "If: T =";T
4010 PRINT "    h =";H
4020 PRINT "    lambda =";L
4030 PRINT "then<ENTER>";:INPUT ":",I$
4040 IF I$="" GOTO 4080
4050 INPUT "       Time (default=1) ";T:IF T=0 THEN T=1
4060 INPUT "              value of h ";H
4070 INPUT "                  lambda ";L:PRINT
4080 INPUT "               n-minimum ";C
4090 INPUT "               n-maximum ";D
4100 INPUT "Multiplier (default=1) ",M:PRINT:IF M=0 THEN M=1
4110 PRINT "MODIFIED POISSON DISTRIBUTION"
4120 PRINT "h =";H
4130 PRINT "lambda =";L
4140 PRINT "T =";T
4150 PRINT "( Multiplier =";M;")":PRINT
4160 J=0:N=C-1:GOSUB 4240:Q=S
4170    FOR N=C TO D
4180    GOSUB 4240
4190    PRINT N;":";S-Q;" ";S:J=J+1
4200    IF J=18 THEN J=0:STOP
4210    Q=S
4220    NEXT N
4230 PRINT O$+O$;"  End P4 ";O$+O$:GOTO 20
4240    IF N<0 THEN S=0:RETURN
```

```
4250     Y=L*(T-N*H):Z=M*EXP(-Y):S=Z
4260     IF N=0 THEN RETURN
4270        FOR K=1 TO N
4280        Z=Z*Y/K:S=S+Z
4290        NEXT K
4300     RETURN
```

P5

```
5000 HOLD=H:LOLD=L:PRINT "T =";T;" is assumed"
5010 PRINT "If h =";H;" then<ENTER>";:INPUT ":",I$
5020 IF I$="" GOTO 5040
5030 INPUT "h-Wert ";H
5040 I=1/(U*T/V-H)
5050 PRINT "corresponding lambda:";I;" if OK, then<ENTER>";:INPUT ":",I$
5060 IF I$="" GOTO 5080
5070 INPUT "lambda ";L
5080 R=0:N=E-1:GOSUB 5180:Q=S
5090   FOR N=E TO F
5100   GOSUB 5180
5110   R=R-A(N-E)*LOG(S-Q)
5120   Q=S
5130   NEXT N
5140 PRINT "-LogLikelihood=";R
5150 INPUT "One more value<ENTER>/ End(otherwise):",I$
5160 IF I$="" GOTO 5030
5170 H=HOLD:L=LOLD:PRINT O$+O$;"  End P5 ";O$+O$:GOTO 20
5180    IF N<0 THEN S=0:RETURN
5190    Y=L*(T-N*H):Z=EXP(-Y):S=Z
5200    IF N=0 THEN RETURN
5210       FOR K=1 TO N
5220       Z=Z*Y/K:S=S+Z
5230       NEXT K
5240    RETURN
```

P6

```
6000 PRINT "Obs: Enter counts per unit time"
6010 PRINT "including also background!"
6020 INPUT "  Source 1 ";A
6030 INPUT "Source 1+2 ";B
6040 INPUT "  Source 2 ";C
6050 INPUT "Background ";D
6060 A=A-D:B=B-D:C=C-D
6070 IF A+C<=B THEN PRINT "h=0":GOTO 6100
6080 K=1/B*(1-SQR(1-B*(A+C-B)/A/C))
6090 PRINT "h=";K:GOTO 6100
6100 PRINT O$+O$;"  End P6 ";O$+O$:GOTO 20
```

P7

```
7000 X=5
7010 PRINT "If: lambda =";L
7020 PRINT "      group limit =";X
7030 PRINT "then<ENTER>";
7040 INPUT ":",I$:IF I$="" GOTO 7070
7050 INPUT "            lambda ";L
7060 INPUT "group limit ";X
7070 PRINT:A=U/EXP(L*T):B=A:K=0:Q=0:D=0:J=1:N=0:IF E=0 THEN N=A(0)
7080 IF B-Q>=X GOTO 7110
7090 K=K+1:A=A*L*T/K:B=B+A:IF E<=K THEN IF K<=F THEN N=N+A(K-E)
```

```
7100 GOTO 7080
7110 IF U-B<=X GOTO 7150
7120 D=D+N^2/(B-Q):PRINT K;":";N;B-Q;B:J=J+1:Q=B:N=0
7130 IF INT(J/19)=J/19 THEN STOP
7140 GOTO 7090
7150 IF K+1>=E THEN IF K+1<=F THEN FOR I=K+1 TO F: N=N+A(I-E):NEXT I
7160 D=D+N^2/(U-Q)-U
7170 PRINT "  ^ :";N;U-Q;U
7180 PRINT "D =";D;J;" groups"
7190 PRINT O$+O$;"  End P7 ";O$+O$:GOTO 20
```

P8

```
8000 X=5
8010 PRINT "Wenn: h =";H
8020 PRINT "       lambda =";L
8030 PRINT "      group limit =";X
8040 PRINT "then<ENTER>";
8050 INPUT ":",I$:IF I$="" GOTO 8090
8060 INPUT "h                       ",H
8070 INPUT "lambda                  ",L
8080 INPUT "group limit ",X
8090 INPUT "k-minimum ";C
8100 K=C:Q=0:D=0:J=1
8110 N=0:IF E<=C THEN FOR I=E TO C:N=N+A(I-E):NEXT I
8120 IF C=0 THEN P=0:B=U/EXP(L*T)
8130 IF C=1 THEN P=U/EXP(L*T):GOSUB 8280
8140 IF C>1 THEN K=C-1:GOSUB 8280:P=B:K=C:GOSUB 8280
8150 IF B-Q>=X GOTO 8200
8160 K=K+1:P=B:GOSUB 8280
8170 IF E<=K THEN IF K<=F THEN N=N+A(K-E)
8180 GOTO 8150
8190 IF U-B<=X GOTO 8230
8200 IF U-B>X THEN D=D+N^2/(B-Q):PRINT K;":";N;B-Q;B:J=J+1:Q=B:N=0
8210 IF U-B>X THEN IF INT(J/19)=J/19 THEN STOP
8220 IF U-B>X GOTO 8160
8230 IF K+1>=E THEN IF K+1<=F THEN FOR I=K+1 TO F:N=N+A(I-E):NEXT I
8240 D=D+N^2/(U-Q)-U
8250 PRINT "  ^ :";N;U-Q;U
8260 PRINT "D =";D;J;" groups"
8270 PRINT O$+O$;"  End P8 ";O$+O$: GOTO 20
8280    Y=L*(T-K*H):Z=U/EXP(Y):B=Z
8290       FOR I=1 TO K
8300       Z=Z*Y/I:B=B+Z
8310       NEXT I
8320    RETURN
```

Printout from a Run of the Programs

```
GW-BASIC 3.22
(C) Copyright Microsoft 1983,1984,1985,1986,1987
60300 Bytes free
Ok
LOAD"P0-8"
Ok
RUN
*********************************** BEGIN **********************************
Choice of Program-Section (1/2/3/4/5/6/7/8) or End<ENTER>:1
********************************** Begin P1**********************************
            min# ? 60
            max# ? 100
time (default=1) ?
 60 ? 3
 61 ? 1
 62 ? 1
 63 ? 4
 64 ? 2
 65 ? 4
 66 ? 4
 67 ? 6
 68 ? 9
 69 ? 11
 70 ? 8
 71 ? 11
 72 ? 17
 73 ? 13
 74 ? 19
 75 ? 19
 76 ? 14
 77 ? 14
 78 ? 16
 79 ? 10
 80 ? 8
 81 ? 10
 82 ? 11
 83 ? 9
 84 ? 5
 85 ? 5
 86 ? 3
 87 ?
 88 ? 3
 89 ? 2
 90 ? 3
 91 ?
 92 ?
 93 ?
 94 ? 3
 95 ? 3
 96 ?
 97 ?
 98 ?
 99 ?
 100 ? 1

              Control(1)/ Correction(2)/ End P1<ENTER>:1

 3 , 1 , 1 , 4 , 2 , 4 , 4 , 6 , 9 , 11 , 8 , 11 , 17 , 13 , 19 , 19 , 14 ,
 14 , 16 , 10 , 8 , 10 , 11 , 9 , 5 , 5 , 3 , 0 , 3 , 2 , 3 , 0 , 0 , 0 , 3 ,
 3 , 0 , 0 , 0 , 0 , 1
Moments (0/1/2):  252  19086  1457602
Observed Intensity (# events per unit time): 75.7381

              Control(1)/ Correction(2)/ End P1<ENTER>:2
```

```
Correction of No.: 95
      true value : 0

              Control(1)/ Correction(2)/ End P1<ENTER>:1

 3 , 1 , 1 , 4 , 2 , 4 , 4 , 6 , 9 , 11 , 8 , 11 , 17 , 13 , 19 , 19 , 14 ,
 14 , 16 , 10 , 8 , 10 , 11 , 9 , 5 , 5 , 3 , 0 , 3 , 2 , 3 , 0 , 0 , 0 , 3 ,
  0 , 0 , 0 , 0 , 0 , 1
Moments (0/1/2):  249  18801  1430527
Observed Intensity (# events per unit time): 75.50603

              Control(1)/ Correction(2)/ End P1<ENTER>:

**********************************  End P1 *********************************
Choice of Program-Section (1/2/3/4/5/6/7/8) or End<ENTER>:2
********************************** Begin P2*********************************
Poisson Distribution(1)/ Binomial Distribution(2):1

If lambda = 75.50603 <ENTER>, else new value of lambda:
                       Multiplier (default=1) ? 100
 Minimal value (prob.* multiplier) by display ? .5
        Max. value of k by display ? 100

POISSON DISTRIBUTION with lambda = 75.50603   (Multiplier = 100 ):

 58 :  .5753236   2.177291
 59 :  .736278   2.913569
 60 :  .9265571   3.840126
 61 :  1.146896   4.987022
 62 :  1.396735   6.383756
 63 :  1.673998   8.057755
 64 :  1.974953   10.03271
 65 :  2.294167   12.32687
 66 :  2.624597   14.95147
 67 :  2.957805   17.90928
 68 :  3.284295   21.19357
 69 :  3.593972   24.78754
 70 :  3.876665   28.66421
 71 :  4.122698   32.7869
 72 :  4.323452   37.11036
 73 :  4.471873   41.58223
 74 :  4.562883   46.14511
 75 :  4.593669   50.73878
continue<ENTER>/ End(any other key):
 76 :  4.563812   55.3026
 77 :  4.475263   59.77786
 78 :  4.332171   64.11003
 79 :  4.14057   68.2506
 80 :  3.907975   72.15857
 81 :  3.642909   75.80148
 82 :  3.35441   79.15588
 83 :  3.051544   82.20743
 84 :  2.742976   84.9504
 85 :  2.436603   87.387
 86 :  2.139281   89.52628
 87 :  1.856651   91.38294
 88 :  1.593049   92.97598
 89 :  1.351515   94.3275
 90 :  1.133861   95.46136
 91 :  .940806   96.40216
 92 :  .7721362   97.17429
 93 :  .6268918   97.80119
```

```
continue<ENTER>/ End(any other key):
 94 :  .5035544   98.30474
**********************************  End P2 **********************************
Choice of Program-Section (1/2/3/4/5/6/7/8) or End<ENTER>:7
********************************** Begin P7**********************************
If: lambda = 75.50603
      group limit = 5
then<ENTER>:

 58 : 0  5.421458  5.421458
 61 : 4  6.996237  12.41769
 63 : 5  7.646134  20.06383
 65 : 6  10.63012  30.69395
 66 : 4  6.535254  37.2292
 67 : 6  7.364941  44.59414
 68 : 9  8.177903  52.77204
 69 : 11  8.948998  61.72104
 70 : 8  9.652908  71.37395
 71 : 11  10.26553  81.63948
 72 : 17  10.7654  92.40488
 73 : 13  11.13497  103.5399
 74 : 19  11.36159  114.9014
 75 : 19  11.43825  126.3397
 76 : 14  11.3639  137.7036
 77 : 14  11.14342  148.847
 78 : 16  10.78711  159.6341
 79 : 10  10.31003  169.9441
Break in 7130
Ok
CONT
 80 : 8  9.730866  179.675
 81 : 10  9.070846  188.7459
 82 : 11  8.352493  197.0983
 83 : 9  7.598358  204.6967
 84 : 5  6.830017  211.5267
 85 : 5  6.067139  217.5939
 86 : 3  5.326813  222.9207
 88 : 3  8.589752  231.5104
 90 : 5  6.188599  237.699
 93 : 0  5.826187  243.5252
  ^ : 4  5.474793  249
D = 42.97592  29  groups
**********************************  End P7 **********************************
Choice of Program-Section (1/2/3/4/5/6/7/8) or End<ENTER>:3
********************************** Begin P3**********************************
 first guess ?
second guess ? .0001

Iteration No. 1 : 0  7862.75
                .0001  7752.875
Iteration No. 2 : .0001  7752.875
                7.156086E-03 -80008
Iteration No. 3 : .0001  7752.875
                7.233411E-04  6927.5
Iteration No. 4 : 7.233411E-04  6927.5
                5.955139E-03 -34007.75
Iteration No. 5 : 7.233411E-04  6927.5
                1.608722E-03  5219.5
Iteration No. 6 : 1.608722E-03  5219.5
                4.314369E-03 -8404.25
Iteration No. 7 : 1.608722E-03  5219.5
                2.645303E-03  2010.75
Iteration No. 8 : 2.645303E-03  2010.75
                3.294872E-03 -1025.25
```

```
continue<ENTER>/ break (any other key)
Iteration No. 9 : 3.294872E-03 -1025.25
                  3.075514E-03  112.625
Iteration No. 10 : 3.075514E-03  112.625
                  3.097226E-03  5.875
Iteration No. 11 : 3.097226E-03  5.875
                  3.098421E-03  0
Iteration No. 12 : 3.098421E-03  0
                  3.098421E-03  0

h= 3.098421E-03              lambda= 98.56533
*********************************** End P3 **********************************
Choice of Program-Section (1/2/3/4/5/6/7/8) or End<ENTER>:4
*********************************** Begin P4**********************************
If: T = 1
    h = 3.098421E-03
    lambda = 98.56533
then<ENTER>:
             n-minimum ? 55
             n-maximum ? 100
Multiplier (default=1) 100

MODIFIED POISSON DISTRIBUTION
h = 3.098421E-03
lambda = 98.56533
T = 1
( Multiplier = 100 )

 55 : 4.511784E-02   .108333
 56 : 7.279341E-02   .1811264
 57 : .1144363   .2955627
 58 : .175336   .4708987
 59 : .2619016   .7328002
 60 : .3814584   1.114259
 61 : .5419002   1.656159
 62 : .7509921   2.407151
 63 : 1.015502   3.422653
 64 : 1.34019   4.762844
 65 : 1.726424   6.489268
 66 : 2.17135   8.660618
 67 : 2.66667   11.32729
 68 : 3.198653   14.52594
 69 : 3.747571   18.27351
 70 : 4.289606   22.56312
 71 : 4.797831   27.36095
 72 : 5.244002   32.60495
Break in 4200
Ok
CONT
 73 : 5.602143   38.20709
 74 : 5.849842   44.05694
 75 : 5.972348   50.02928
 76 : 5.961331   55.99061
 77 : 5.818871   61.80949
 78 : 5.554478   67.36396
 79 : 5.186165   72.55013
 80 : 4.736412   77.28654
 81 : 4.231209   81.51775
 82 : 3.698593   85.21634
 83 : 3.163452   88.37979
 84 : 2.64711   91.0269
 85 : 2.167618   93.19452
 86 : 1.737579   94.9321
 87 : 1.362534   96.29463
```

```
 88 : 1.046516   97.34115
 89 : .7860641   98.12721
 90 : .5787048   98.70592
Break in 4200
Ok
CONT
 91 : .4169617   99.12288
 92 : .2934342   99.41631
 93 : .2030487   99.61936
 94 : .1372681   99.75663
 95 : 9.039307E-02   99.84702
 96 : 5.895996E-02   99.90598
 97 : 3.760529E-02   99.94359
 98 : 2.272034E-02   99.96631
 99 : 1.441193E-02   99.98072
 100 : 8.033753E-03   99.98876
*********************************** End P4 ***********************************
Choice of Program-Section (1/2/3/4/5/6/7/8) or End<ENTER>:8
*********************************** Begin P8***********************************
Wenn: h = 3.098421E-03
      lambda = 98.56533
      group limit = 5
then<ENTER>:
k-minimum ? 60
 62 : 5  5.993805  5.993805
 64 : 6  5.865671  11.85948
 66 : 8  9.705481  21.56496
 67 : 6  6.639977  28.20493
 68 : 9  7.964661  36.1696
 69 : 11  9.331451  45.50105
 70 : 8  10.68111  56.18216
 71 : 11  11.94659  68.12876
 72 : 17  13.05758  81.18633
 73 : 13  13.9493  95.13562
 74 : 19  14.56609  109.7017
 75 : 19  14.87122  124.573
 76 : 14  14.84368  139.4166
 77 : 14  14.48895  153.9056
 78 : 16  13.83061  167.7362
 79 : 10  12.91367  180.6499
 80 : 8  11.79369  192.4435
 81 : 10  10.53554  202.9791
Break in 8210
Ok
CONT
 82 : 11  9.209518  212.1886
 83 : 9  7.87706  220.0657
 84 : 5  6.591248  226.6569
 85 : 5  5.397476  232.0544
 87 : 3  7.719132  239.7735
  ^ : 12  9.226486  249
D = 12.41113  24  groups
*********************************** End P8 ***********************************
Choice of Program-Section (1/2/3/4/5/6/7/8) or End<ENTER>:5
*********************************** Begin P5***********************************
T = 1  is assumed
If h = 3.098421E-03  then<ENTER>:
corresponding lambda: 98.56533  if OK, then<ENTER>:
-LogLikelihood= 823.8885
One more value<ENTER>/ End(otherwise):
h-Wert ? .0032
corresponding lambda: 99.56216  if OK, then<ENTER>:
-LogLikelihood= 824.7231
One more value<ENTER>/ End(otherwise):
```

```
h-Wert ? .0030
corresponding lambda: 97.61834  if OK, then<ENTER>:
-LogLikelihood= 824.8256
One more value<ENTER>/ End(otherwise):
h-Wert ? 3.098421E-03
corresponding lambda: 98.56533  if OK, then<ENTER>:x
lambda ? 99
-LogLikelihood= 824.0572
One more value<ENTER>/ End(otherwise):
h-Wert ? 3.098421E-03
corresponding lambda: 98.56533  if OK, then<ENTER>:x
lambda ? 98
-LogLikelihood= 824.0807
One more value<ENTER>/ End(otherwise):x
*********************************  End P5 **********************************
Choice of Program-Section (1/2/3/4/5/6/7/8) or End<ENTER>:6
********************************* Begin P6**********************************
Obs: Enter counts per unit time
including also background!
  Source 1 ? 383.67
Source 1+2 ? 445.57
  Source 2 ? 371.97
Background ? 2.49
h= 1.851007E-03
*********************************  End P6 **********************************
Choice of Program-Section (1/2/3/4/5/6/7/8) or End<ENTER>:
sure<ENTER>/ escape(any other key):
*********************************   END    *********************************
Ok
system
```

REFERENCES

1. Andersen,A.H., N.Keiding et al., *Opgavesamling i anvendt statistik*, Afdeling for teoretisk statistik, Aarhus Universitet[1] (1976).
2. Bohr, N., *Atomic Physics and Human Knowledge*, John Wiley and Sons (1958).
3. Bortkiewicz, L. von, *Das Gesetz der kleinen Zahlen*, Springer (1898).
4. Bortkiewicz, L. von:, *Die radioaktive Strahlung als Gegenstand wahrscheinlichkeitstheoretischer Untersuchungen*, Springer (1913).
5. Boyer, C.B. et al. (ed.), *Dictionary of Scientific Biography*, Scribners (1970–78).
6. Brockmeyer, E., H.L. Halstrøm and A. Jensen, *The Life and Works of A.K. Erlang*, Acta Polytechnica Scandinavica (1960).
7. Chadwick, J. (scientific adviser), *The Collected Papers of Lord Rutherford of Nelson*, Georg Allen and Unwin **I,II** (1962-63).
8. Evans, R.D., *The atomic nucleus*, McGraw–Hill[2] (1955).
9. Faure, G., *Principles of isotope geology*, Wiley (1977).
10. Feller, W., *An Introduction to Probability Theory and its Applications*, John Wiley and Sons **(I,**2nd ed.**)**, **(II)** (1957,1966).
11. Hald, A., *Statistical Theory with Engineering Applications*, John Wiley and Sons (1960), (first published 1948).
12. Johansen, A., L. Sarholt-Kristensen and K.G. Hansen, *Risikomomenter ved anvendelse af ioniserende stråling i fysikundervisningen*[2], H.C.Ørsted Institutet (1980).
13. Obel, S., K. Kristensen and K. Heydorn, *Experimentel kernefysik*[2], Gyldendal (1970).
14. Pearson, E.S. and J. Wishart (ed.), *"Student's" Collected Papers*, Cambridge University Press (1958).
15. Waerden, B.L. van der (ed.), *Sources of quantum mechanics*, Dover (1965).

[1]This is a collection of exercises, mainly related to data from statistical experiments. No special substitute in English is suggested for this edition.

[2]For this edition we recommend to replace the books listed under the references no. 8,12 and 13 by Herman Cember: Introduction to Health Physics, 2nd.ed., Pergamon Press (1983).

Index

A

accidents, 142, 150
activity, 136ff
age
  determination of, 139, 153
  of earth, 106
  of rocks, 153
  of universe, 139
alpha-radiation/particle, 104, 106
approximating polynomial, 124
approximation, Poisson, 53ff
arrival time, 2, 36
atomic model, 106ff, 108
attenuation, 77, 144
attenuation constant, 140
average, 12, 129

B

bacteria, 144
bacteriophage, 158
Becquerel, H., 104
Bernoulli, J., 93
Bernoulli variable, 23
beta-radiation/particle, 104
binomial coefficient, 24
Bohr, N., 108ff
bombs over London, 141
Bortkiewicz, 55, 99, 101, 129

C

cascade process, 137ff
  with branching, 151
Cesium-Barium experiment, 154
Challenger catastrophe, 12
Chebyshev, 95
Chebyshev's inequality, 120
chi-square test, 81, 104, 164ff
convergence
  of sequence, 122
  of series, 123
cosmic radiation, 77, 146
Curie, M. and P., 104

D

dead time, 49, 73ff
  determination of, 132ff, 163ff
decay, 1, 27

spontaneous, 28
decay constant, 135
de Moivre, A., 94ff
density, 115
density function, 116, 119
of chi-square distribution, 82
detection probability, 49, 125
discrepancy, 86ff, 165
distribution, 14ff
Bernoulli, 23
binomial, 23ff, 53ff, 126, 162
chi-square, 81
continuous, 115
empirical, 70, 84
exponential, 37, 121
geometric, 114
normal, 57
Poisson, 36, 53ff, 127, 162
mixtures of, 130
modified, 73, 164
spatially uniform, 59ff, 127
support of, 115
uniform, 116
distribution function, 15, 117
empirical, 70

E

Einstein, A., 109ff
Erlang, A.K., 102, 103
estimator, 112
central, 129
maximum likelihood, 71, 74ff
events, with low probability, 12
excited state, 27
expectation, 10, 12
extinguisher, 78

F

fact, empirical, 12
Feynman, R.P., 12
Fisher, R.A., 104
fractile, 84
frequency, 10
cumulative, 70
interpretation, 11
relative, 10
stabilisation of, 12ff

G

Geiger, H.W., 40, 105
Geiger-Müller tube, 73, 78, 105
goodness of fit, 89
Gosset, W.S., 59, 100, 101
ground state, 27
group limit, 92
grouping, 86, 164
Grundbegriffe, 98

H

haemacytometer, counts with, 59
half-distance, 140
half-life, 104, 135
stochastic, 136
histogram, 41
horse kicks, 55

I

independence, 21ff, 113, 119
insurance, 142
intensity, 29ff
in point process, 62
observed, 40
registered, 71, 131

J

jaundice, 146

K

Kelvin, Lord, 106
Kolmogorov, A.N., 97ff

L

Laplace, 95
law
  of large numbers, 10, 93, 95, 98, 120
  of stabilisation, 12
  of total probability, 125
leukemia, 146
Lexis, W., 129
Lexis coefficient, 129
likelihood function, 74, 122
Lundberg, F., 102
lyses, 158

M

maximum likelihood, 74, 104, 121ff, 130
mean value, 10, 15, 116ff
model
  building of, 5ff
  stochastic, 9ff, 13
moments, 162
mutation, 144

N

nuclear charge, 152

P

paradox, 125, 128
paralysis, 49ff
Pearson, K., 104
phenomena
  deterministic, 9
  stochastic, 9
plaque, 158
point process, 62
Poisson, S.-D., 95ff
  approximation, 53ff
distribution, 36
  mixtures of, 130, 151
  process, 36, 51, 101ff
    superposition of, 47ff, 68
probability, 10ff
  a posteriori, 93
  a priori, 11
  Cardano-Fermat-Pascal, 93
  conditional, 22
  space, 14, 113
  success, 23

Q

quantum theory, 110
quark model, 139
queue, 146

R

radiation background, 49, 66
radioactive
  decay, models of, 36, 131, 133ff, 136ff
  series, 48
  source, 66
    from RISO, 66ff
radioactivity, 1ff, 27, 104, 134
  daughter process, 67
  mother process, 67
random variable, 13ff
  discrete, 15
  identically distributed, 15
    sequence of independent, 22
regularity condition, 33ff, 125
rubidium-strontium method, 153
Rutherford, E., 40, 104ff
Rutherford-Geiger experiment, 40ff, 48, 79, 101, 106ff, 146

S

sample, 69, 83, 121
sample average, 129
sample space, 14, 45
sample variance, 129
sclerosis, 157
series, infinite, 123ff
  geometric, 123
significance level, 88
spontaneity, 52, 69, 75, 107, 121

standard deviation, 119
statistical thermodynamics, 10
stochastic, 9
  process, 36
stopping time, 19
storage, of data, 161
Student, 59
submarine, 151
success variable, 23, 32

T

Taylor series,124
telephony, 101ff, 150
time invariance, 17ff
traffic count, 144
trials, 10
  Bernoulli, 23
two-source method, 66, 78, 132ff

V

variance, 119

W

waiting time, 2, 37
  global, 135
  individual, 134
  mean, 30, 37
  paradox, 125, 128
  population, 135
water fleas, 142
World Cup, 156

Y

yeast cells, 59ff